ANCIENT CHINESE TIMBER ARCHITECTURE

www.royalcollins.com

A Journey through History

ANCIENT CHINESE TIMBER ARCHITECTURE

By Liu Haibo

Translated by Jianning Ding, Yang Yang, and Xuemeng Angela Li

Books Beyond Boundaries

ROYAL COLLINS

Ancient Chinese Timber Architecture: A Journey through History

By Liu Haibo
Translated by Jianning Ding, Yang Yang, and Xuemeng Angela Li

First published in 2025 by Royal Collins Publishing Group Inc.
Groupe Publication Royal Collins Inc.
550-555 boul. René-Lévesque O Montréal (Québec) H2Z1B1 Canada

10 9 8 7 6 5 4 3 2 1

ISBN: 978-1-4878-1324-6

To find out more about our publications, please visit www.royalcollins.com.

CONTENTS

FOREWORD

The writing of this book was motivated by a long-standing perplexity regarding the significance of ancient Chinese architecture. This architectural style not only serves as a vital conduit for Chinese culture but also represents a profound art form that encapsulates the essence of traditional Chinese philosophy, culture, and artistic aesthetics. Given its cultural, ideological, and artistic importance, ancient architecture should be regarded as a crucial component for fostering art appreciation, enhancing public intellectualism, and advancing social civilization.

In reality, the public's understanding and appreciation of the artistry of ancient Chinese architecture remains limited. This often results in only superficial admiration based on intuitive perception. In contrast to the appreciation of ancient Western architecture, where a general understanding of its genres and artistic styles is sufficient, our engagement with the art of ancient architecture should extend beyond mere sensual delight. As inheritors of our cultural heritage and custodians of our history, we are obliged to cultivate a deeper cognitive understanding. Such an endeavor not only reflects the advancement of social civilization but also lays the foundation for social progress.

There is currently a noticeable gap in the public's understanding of the artistry of ancient architecture. While professional academic research may be challenging for the layperson to comprehend, the superficial introductions provided in travel guides fail to foster a true appreciation for the art of ancient architecture. What is required is a conduit that facilitates the integration of specialized academic research with public knowledge, a point where the two domains can be effectively unified. Even a basic understanding and appreciation should be accompanied by fundamental and systematic knowledge. For example, while we may be impressed by the ingenuity of bracket sets, we should also investigate the intricacies of their craftsmanship, the diverse functions of brackets employed in various contexts, and the distinctions between Tang-Song and Ming-Qing bracket styles. Such an in-depth understanding is essential for an authentic appreciation of the art of ancient architecture, and it serves as the primary impetus for writing this book.

The selection of content and the determination of the depth of analysis in the field of ancient architecture, as well as the decision regarding the narrative structure, represent the most challenging aspects of the writing process. First and foremost, the text should be concise and straightforward, with minimal use of words to ensure clarity and comprehension. However, the challenge of conciseness is to ensure both overall understanding and deep comprehension. Secondly, I hope that the text will be accessible to a general audience, employing an informal and engaging style of language and a narrative style that is light and captivating. Nevertheless, this description style necessitates a higher level of logical coherence and conceptual structure. Furthermore, I hope that the narrative will diverge from the conventional chronological or biographical structure, unfolding instead in the manner of a cinematic narrative characterized by dramatic peaks and troughs.

The depth of one's research into the art of ancient architecture ultimately determines the extent to which one can appreciate the nuances of these works. Due to my limited knowledge, the current book does not adhere to either the academic or the popular style. It is neither profound nor simplistic. Furthermore, even focusing on ancient timber architecture presents a significant challenge due to the complexity of ancient architectural systems and the vast array of architectural forms. The original intention of creating a concise and compact book had to be revised as the writing progressed.

This approach enables the original intention of finding some point of convergence between popular knowledge and academic scholarship to be realized. Nevertheless, I hope that the general public's understanding of the art of ancient architecture will be enhanced.

Upon encountering ancient Chinese architecture once more, one not only acknowledges its aesthetic appeal but also experiences a sense of tranquility and immersion in the multifaceted beauty expressed in form, structure, and decoration. This is the essence of this book.

CHAPTER 1

FIRST ENCOUNTER WITH ANCIENT ARCHITECTURE

The official residence of the President of the National Government, also known as Meiling Palace, located in Zhongshan Mountain Nanjing, is a modern building built in 1931 (the 20th year of the Republic of China).

What Is Ancient Architecture?

Architecture is a loanword originally expressed in Chinese as 'house.' When discussing ancient Chinese architecture, iconic landmarks such as the Forbidden City, the Imperial Ancestral Temple, the Temple of Heaven in Beijing, and ancient dwellings and temples often come to mind. According to Mr. Liang Sicheng, all Chinese architecture prior to the establishment of the Republic of China in 1912 is considered to be ancient architecture.[1] While this classification may be subject to modifications as history progresses, it is currently widely accepted. Regarding buildings constructed after 1912 that possess all the characteristics of Chinese architecture, they are often labeled as traditional or replica architecture. During the Republic of China era, a significant number of high-quality buildings were constructed by incorporating and drawing inspiration from Western architecture while still maintaining the essence of Chinese architecture. These constructions are commonly referred to as Republican or modern architecture.

It is challenging to make generalizations about ancient Chinese architecture, given the vast geographical expanse of China and the diverse ethnic groups, variations in geography, natural environment, climate, available resources, cultural practices, religious beliefs, aesthetic traditions, and regional differences in architectural materials and techniques. Consequently, unique architectural needs and conditions have emerged, leading to the emergence of diverse architectural styles and forms across different ethnicities and regions. Despite the distinct characteristics of architectural styles in different regions and ethnic groups, there is a common lineage and significant similarities in terms of architectural systems. They share a common foundational framework but exhibit different styles within a unified architectural system.[2]

FOLLOWING PAGE
Residential houses in Dong Village, Zhaoxing Town, Guizhou Province.

仁團花橋

穆師傅銀飾坊

TOP Residential houses in Huangling Village, Wuyuan County, Jiangxi Province.

BOTTOM Residential houses in earth buildings, Yongding District, Longyan City, Fujian Province.

Eastern Architecture and Western Architecture

It is widely acknowledged within the academic community that world architecture can be categorized into two major groups: Western architecture and Eastern architecture. Western architecture is primarily based on European architecture. Europeans hold the belief that Europe has played a pivotal role in shaping the culture, technology, and history of modern world civilization. Likewise, it is believed that Western architecture continues to influence the development of architecture in countries around the world. This influence is often not attributed to other architectural systems.[3]

The "Tree of Architecture" in *A History of Architecture on the Comparative Method* by B. Fletcher, UK.

Eastern architecture is comprised of three principal systems. The Indian architecture of the South Asia region has a long history and features impressive buildings such as the Great Stupa at Sanchi. Islamic architecture originated in West Asia (Muslim architecture) and spread throughout Europe, Asia, and Africa, exerting extensive influence around the world like its counterparts. Chinese architecture has the longest history and the most unified style among Eastern architecture. It is also known for its remarkable characteristics. The architectural styles of Japan, Korea, and the Indo-China Peninsula have all been steadily integrated into this system for a long time.[4]

As Westerners believed, Western architecture has played a significant role in the advancement of modern architecture, with notable achievements and contributions. While Western architecture is a prominent influence, the development of world architecture is a collaborative effort of all peoples around the globe." Modern civilization is the result of the evolution of human society. As societies become more advanced, the need for cultural exchange increases."[5] The Japanese architect Ito Chuta had already criticized Europe's lack of understanding of oriental culture and art.[6] British scholar Joseph Needham also acknowledged this issue, stating that "this vast and prosperous civilization at the other end of Eurasia is at least as complex and varied as their own."[7]

These attitudes and views have undergone significant transformation in line with the development of the world.

On the one hand, Westerners have gained more knowledge about the world beyond their own, with a deeper understanding of the significance of embracing cultures from all regions and peoples around the world to promote the development of science and culture for all of humanity. There has been a notable increase in the number of research papers published on Chinese architecture. The practice of architecture worldwide, with the incorporation of Chinese architectural expertise, has significantly advanced the development of global architecture.

On the other hand, the awakening and attention from within East Asia has enhanced the influence of Eastern architecture. Around 1930, a group of overseas students of architecture majors, represented by Lü Yanzhi, Liang Sicheng, Liu Dunzhen, Yang Tingbao, and Tong Li, received systematic training in Western architecture and examined and studied Chinese architecture with international visions and modern scientific methods. This approach enabled them to grasp the academic development of Chinese architecture more accurately and systematically.

Since 1937, Liang Sicheng and Lin Huiyin have conducted extensive research, successively traveling through more than 200 counties in 15 provinces of China, surveying and photographing over 2,000 ancient architectural relics from the Tang, Song, Liao, Jin, Yuan, Ming, and Qing dynasties, including the Liao Dynasty Guanyin Pavilion of Dule Temple in Tianjin Ji County, the Liao Dynasty Guangji Temple in Baodi District of Tianjin, the Liao Dynasty Longxing Temple in Zhengding, Hebei Province, the Liao Dynasty Wooden Pagoda in Ying County, Shanxi Province, the Liao Dynasty Huayan Temple and Shanhua Temple in Datong, and the Sui Dynasty Anji Bridge in Zhaozhou, Hebei Province. The findings of these important investigations were disseminated internationally, garnering international interest. Liang Sicheng's *A Pictorial History of Chinese Architecture*, written in English during the period of the War of Resistance against Japanese Aggression, played a pivotal role in introducing Chinese architecture to the global community. Liang Sicheng's academic achievements were also highly praised by foreign academic communities. Joseph Needham once asserted that Liang Sicheng was a "master of Chinese architectural history."

Meanwhile, Japanese architects such as Tadashi Sekino, Ito Chuta, and Daijo Tokiwa, who undertook numerous trips to China for architectural field research, meticulously documented the preservation status of Chinese architectural heritage in the early 20th century. They notably employed archaeological methodologies to conduct thorough investigations into architectural structures and published several research works. These endeavors significantly contributed to the broader recognition and global influence of Chinese architecture.

Why Timber Framing?

The process of architectural formation is intricate, reflecting diverse social practices. Chinese and Western civilizations, in particular, evoke distinct impressions through their architecture. Ancient Greek temples and medieval European

LEFT In 1902, the Japanese scholar Ito Chuta (front left) inspected architecture in Guizhou (at the home of a local squire).

RIGHT In 1935, Liang Sicheng and Lin Huiyin on the roof of the Hall of Prayer for Good Harvest.

churches are notable for their grandeur and imposing presence, while the Imperial Ancestral Temple and the Forbidden City exemplify a meticulously structured layout. Among the architectural systems observed globally, only China and its neighboring regions, including Korea and Japan, predominantly utilize wood as the primary material for residential structures, employing the timber frame system. In contrast, Western, Indian, and Eastern Islamic architectural traditions largely adopt masonry structural systems. Timber framing represents a distinctive feature that distinguishes Chinese architecture from other architectural traditions.

China, despite its plentiful stone resources, adopted timber framing as the primary architectural system. Conversely, blessed with ample wood, the Western world predominantly embraced masonry architectural systems. This prompts the question of why these choices deviated from the apparent suitability of the available materials. The answer to this question cannot be attributed to mere environmental factors or cultural and philosophical distinctions.

In addressing this inquiry, Mr. Liang Sicheng offered a comprehensive overview: "Chinese architecture primarily relies on wooden frames, which results in houses with comparatively less durability due to the inherent nature of wood. However, the fundamental rationale for this choice stems from the Chinese cultural perspective, which values non-attachment to the enduring physical presence of objects."[8] Mr. Liang believes that wooden frames were favored because of the Chinese cultural inclination toward non-attachment to physical forms. While aligning with the notion of following the natural lifespan of objects, it's crucial to acknowledge that the Chinese also aspired for eternity. They might have preferred to maintain flexibility in their choices regarding possessions. However, what choices were made and for what reasons? This question has drawn increased attention from architectural scholars, who have offered additional insights following Mr. Liang's initial assertion.

In his book *Types and Structures of Chinese Architecture*, Mr. Liu Zhiping offers insights on the prevalence of timber frames in Chinese architecture: "The earliest settlements in China are primarily located in loess regions like the Central Plains, where wood resources were more abundant than stone. Consequently, stone construction is not a common feature. Wood, being lightweight and durable, provided tactile comfort and was easily accessible for construction. Moreover, the post-and-beam structure enabled the installation of doors and windows with ease. These factors contributed to the widespread adoption of timber framing in the palatial structures of early primitive Chinese society."[9] Mr. Liu Zhiping concludes that the preference for timber framing was primarily influenced by environmental factors.

In his work *Principles of Chinese Classical Architectural Design*, Mr. Li Yunhe offered a different perspective in response to Liu Zhiping's view and developed his conclusion by referencing the insights of Joseph Needham. He stated, "It is inaccurate to claim that China lacks the stone resources necessary for building large-scale structures akin to those in Europe and West Asia. Rather than using stone for large-scale structures, the Chinese employed it for mausoleums, marble columns, and monuments, often incorporating details from classical woodwork. Furthermore, stone was employed in the construction of pavements, courtyards, and roadside paths."[10] Li Yunhe thus suggests that "objective factors such as the environmental conditions and geographical constraints were not the primary factors influencing the adoption and evolution of timber framing in China. In essence, the availability of wood resources was not ubiquitous throughout China."[11]

Li Yunhe also challenged Joseph Needham's assertion that Chinese slavery and the adoption of timber framing in Chinese architecture were closely intertwined, "A better understanding of Chinese social and economic conditions might provide a more accurate insight. It appears that the form of slavery prevalent in Western cultures never existed in parallel within China. Western nations historically deployed large numbers of men to undertake strenuous labor in stone factories. For example, the colossal sculptures characteristic of Assyria or Egypt have no equivalent in Chinese culture. The scale of such endeavors necessitates vast amounts of labor for the transportation and sculpting of massive rocks. While Qin Shi Huang's (First Emperor of Qin, 259–210 BC) absolute authority during the construction of the earliest Great Wall is unprecedented, it's undeniable that ancient and medieval China could mobilize significant labor forces. However, the fundamental character of Chinese architecture was already established by then. Therefore, it can be reasonably assumed that the use of timber framing contributed to the absence of a large population of slaves.[12] Li Yunhe references Guo Moruo's *The Slavery Era*, citing statements such as "There was undoubtedly a large number of slaves in the Yin Dynasty" and "The Yin people owned many royal slaves and a considerable amount of private slaves as well." He asserts that "social systems and productivity were not the primary drivers behind the use of timber frames in buildings. Ancient Chinese rulers had the ability to mobilize a large scale of labor forces.[13]

Li Yunhe posits that theories predicated on objective conditions are susceptible to close scrutiny and serve as a compelling argument for the prevalence of timber framing in Chinese architecture. Li Yunhe concludes: "The enduring use of timber-frame hybrid structures in Chinese architecture is rooted in their recognition as the most rational construction method, established through a process of selection and testing."[14] Until then, Li Yunhe had approached the heart of the matter. Yet, I believe there remains room for further exploration to fully uncover the core of the issue.

In *Discussions on Chinese Architecture*, Mr. Zhang Jiaji comments on the issue of the adoption of timber framing in Chinese architecture: "In fact, this issue, despite its significant implications within architectural history, has not attracted the attention of the Chinese architectural community. Not a single scholarly paper has delved into the study of it. Existing literature on Chinese architecture merely offers vague and inconsistent discussions on the matter, lacking substantial depth or logical coherence. Consequently, the historiography of Chinese architecture is rife with speculative narratives, largely unchallenged and unsupported by empirical evidence."[15] He praises Li Yunhe for pioneering the inquiry into why Chinese architecture uses timber frames in a thematic review. However, he also suggests that his proposed answer may not provide a wholly persuasive explanation.

Zhang Jiaji focused on summarizing and analyzing Liu Zhiping's theory of natural conditions, architect Xu Jingzhi's theory of economic level, British scholar Joseph Needham's theory of the social system, and Li Yunhe's theory of technical standards. He also set out the reasons why he disagreed with each theory and why timber-frame construction should be adopted. Unfortunately, Zhang Jiaji also did not develop a more mature and complete understanding of his own position.

To a certain extent, the scholars mentioned above attempted to identify a starting point to analyze the adoption of timber frames in Chinese architecture.

They shared a focus on the subject of the question, namely, the reasons behind the use of timber frames in Chinese architecture. If one seeks to understand the answer within the context of Chinese architecture itself, it is natural to consider the social system, natural conditions, economic level, and technological development that created it. These are undoubtedly the content and methodologies that relevant research must prioritize. However, we can also analyze this issue within the context of global architectural systems. If we consider the Western architectural system from a Western perspective and attempt to prove why Western architecture mainly consists of stone structures instead of wood, can this provide a breakthrough in understanding the adoption of timber frames in Chinese architecture? Alternatively, we can explore this issue by comparing the origins of Eastern and Western architecture.

The geographical environment is a key factor in the emergence of ancient man. It requires a relatively mild climate, easy access to food, and a guarantee of safety. Once these basic conditions for survival were in place, ancient man would also have had a shelter to protect him from the biting winds of freezing winter and the burning sun of scorching summer. It is conceivable that ancient humans initially utilized only natural conditions such as caves, hollows, crypts, and nests (or, according to Darwin's theory of evolution, they were able to use natural habitats even before they officially became human). As long as a place provides a sense of shelter, it can be considered a habitat. This process allows for the possibility arises of continuously maintaining existing habitats and gradually imitating nature to create new ones. The first challenge faced by ancient humans was the selection of appropriate building materials, given the limitations of natural geological resources and the availability of convenient, accidental options. They could excavate caves in the hillside areas and holes in the earth, layer branches in trees, or even construct caves with stones. Under their respective natural, geological, and material conditions, the first human inhabitants of each region completed the development of their selection of building materials, formed a common understanding, and also developed their cognitive, aesthetic, habitual, and technical genes based on the subsequent experience they accumulated through selection elimination.

Ancient Egyptian and Western Asian architecture is regarded as the inaugural stage of the Western architectural system. In the Mediterranean, the eastern mountain ranges along the coast are constructed using stone. There is also a plentiful supply of wood and earth, although stone is much more abundant in the long run. In mountainous areas, natural caves were evidently the first choice for our ancestors as a reference for building their own dwellings. It is also not surprising that any material would be used as long as it could be used for building. Therefore, earth and wood materials were widely used in the early historical period, even in the Mediterranean region. In terms of functionality, stone and earth are relatively suitable for building enclosures, while wood is clearly more advantageous for building roofs.

From a structural maintenance perspective, stone and earth each have their own advantages and disadvantages. Stone is a challenging material, requiring significant time and effort to process and finish. Its inherent hardness and bulkiness present unique challenges in construction. Conversely, stone offers the significant advantage of durability and strength, which protects it from collapsing due to weathering and erosion. In contrast, earth is a more straightforward material to work with but is more susceptible to damage.

Under the combined influence of the availability and characteristics of different materials, people in the Mediterranean region gradually adopted stone. It may have started with one tribe, then gradually prevailed and influenced others until it became popular. The choice of stone is both the result of natural conditions and of preferences and habits formed over time, which then become the basis for the perception of housing structures. It is both natural and social selection. In fact, similar transitions from accidental natural selection to inevitable social selection have occurred in other different regions of the world. In many mountainous regions of China, for example, there have long been large numbers of stone houses. It is noticeable that the regions where these stone houses exist are not the main areas of origin of Chinese civilization. Rather, they are either small habitats or migrations in later generations whose building structure is linked to the available regional materials. However, as these structures were mainly based on copying wooden architecture, they were not able to influence wooden architecture to become the mainstream style.

Let's shift the focus back to the upper reaches of the Yellow River region, where Chinese civilization was born. The provinces of Gansu and Shaanxi are almost covered with yellow earth, even on their mountains. "Ancient men rest in caves and live in the wilderness." Habitat construction, which evolved from caves, naturally uses earth. Similarly, the use of wood to build roofs becomes an inevitable choice. As a result, it is natural that "earth" and "wood" develop together as a choice of material. This is why Chinese architecture has a traditional term called "earth-wood construction." To this day, caves and crypt dwellings still exist in China, which fully illustrates our ancestors' choice of building materials. Cave dwellings are more demanding in terms of geological conditions. Regions that do not meet these requirements cannot develop cave structures, which also explains the absence of caves in other regions.

The choice of timber construction in China is not due to the lack of natural resources such as stone, the adoption of relatively simple processing technologies, or the lack of a large labor force to build with stone. All of these are certainly counter-evidence from later generations. The construction of houses did not start all over China, but only in the cradle of Chinese civilization during a certain period in ancient times. The influence of all subsequent conditions cannot change the genes since their initial establishment.

Ever since our ancestors began to build their habitats, their initial choice of materials for copying cave structures has been wood and earth. This is an initial choice of technology, an initial habit, and an initial construction concept. From an initial preferred choice of technology, an initial habit to social consciousness, such a choice has gradually developed into the social consciousness, aesthetic, and ideological outlook that has led to the use of timber framing in Chinese architecture. This kind of deep-rooted awareness and consciousness has continued into modern times. This is how we arrived at the above statement.

Ideological Influence on Architectural Design

In ancient China, architecture was seen as a mere technique for "building houses." A complete academic system for the study of architecture was not established. There were only a few specialized works on the study of architectural ideologies, including those on architectural design and construction.

Of these, *Kao Gong Ji* (*Records of Examination of Craftsman*), *Yingzao Fashi* (*State Building Standards*), and *Gongcheng Zuofa Zeli* (*Manual of Engineering

The Colosseum in Rome (AD 72–80).

Practice) are the three best-known. In essence, these three books are not monographs of academic architectural research but rather architectural codes issued by the imperial court, which are equivalent to today's building codes and regulations. Other famous books include *Mu Jing* (*Wood Classic*) and *Lu Ban Jing* (*Lu Ban Classic*), which are essentially summaries of the technical experience of craftsmen. Joseph Needham quotes Paul Demiéville, the French sinologist and renowned scholar of Tunhuangology, "This is probably because that Confucians did not consider the practice of architecture a suitable occupation for Confucian disciples. As a result, works on architecture are relatively rare in Chinese literature."[16]

This does not mean, however, that there is no theory of Chinese architectural ideologies.

On the contrary, Chinese architectural ideologies were mostly derived from the Chinese philosophical ideology system, which led to the formation of their complicated character. Most analyses of architectural ideologies don't focus on the issues of architectural structures but use them to illustrate social, political, and ideological meanings. Similarly, the ideas of some classical book series on Confucianism and history have often been cited to illustrate the basis of certain architectural structures. The idea of classical ideologies was not only the guideline of action and behavior in ancient China but also the basis of Chinese architectural ideology.

This is particularly the case from the 5th to the 3rd centuries BC during the Spring and Autumn and Warring States periods and the Qin and Han dynasties. This was an important period in the development of the ideological system and viewpoints of Chinese philosophy. It is also the source of Chinese architectural ideologies, which have been constantly quoted, improved and developed by later generations. This is what makes Chinese architectural ideology unique.

Ancient Chinese ideologies and philosophies originated in the social sciences. It mainly discusses social and political issues, ethics, and morality, focusing on man's relationship with nature, society, and himself.[17] The great philosophers of ancient Chinese society, such as Confucius, Mencius, Lao Tsu, and Zhuangzi, were also politicians, educators, and writers, which makes Chinese philosophy fundamentally different from Western philosophy. Western philosophy comes from the natural sciences, which study how nature works.

All human wisdom, such as astronomy, geography, physics, and chemistry, is considered part of philosophy. The great ancient Western philosophers, such as Pythagoras, Democritus, and Aristotle, were also natural scientists.

The focus on nature and science in Western philosophy and the core of society and politics in Chinese philosophy is precisely the reason why the two worlds have followed very different paths in the development of their respective architectural ideologies. As a socio-political and ethical code of conduct, Chinese philosophical ideology is sensual and straightforward, with edifying characteristics or effects. Chinese architectural ideology is the integration of several schools of ideologies and doctrines, among which any ideas that can be used as guidance for architectural behavior will be cited by later generations as their own architectural views to support their architectural understanding and choice of architectural behavior.

Each school of classical ideology has long held different views on its understanding of architecture. Although this understanding is intended to discuss social views and ethics, it has shaped different views of architecture after being reflected in the guiding architectural ideologies of later generations. After a comprehensive review of the ideas of Chinese architecture, one will find that Chinese architecture develops through the fusion and exchange between the contradiction, unification, and communication of the various architectural ideologies of different ideologies and doctrines.

Among these, the conflict and unity between the ideas of "humble palace chamber" and "grand palaces and exquisite pavilions," "great strength" and "appropriate form," and the concept of the "middle way" are particularly significant. Although these were social and ethical choices for royal palaces, they also influenced the general value judgments of both official and popular architecture. This includes the fundamental choice between luxury and simplicity.

The phrase "humble palace chamber" first appeared in Confucius' commentary on Da Yu. He said, "I can find no fault in Yu's character. He had meager food and drink, yet he showed filial piety to the spirits and gods. He wore simple clothes, but displayed the utmost elegance in his sacrificial cap and apron. He lived in a humble palace chamber, but spent all his strength on the moats and waterways. I cannot find a single fault with Yu." The comment basically shows Confucius' approval of Yu's character, which was to eat a simple daily diet, but to serve the ancestors and gods with abundant food; to dress in a simple style, but to wear the hats and ornaments that indicate social status; to live in humble and low houses, and to devote financial, material and labor resources to water conservation projects that support agricultural production.

The idea of frugality reflected in the palace architecture was emphasized and quoted by later generations as the benevolent virtue and governing ethic of the emperor. On the one hand, the idea was used by the emperor himself as a means of self-expression. On the other hand, it was also suggested by the ministers to maintain the image of the emperor and the prosperity of the country by limiting the extravagant behavior of the emperor.

Image of Emperor Wen of Han.

It is recorded in the history of the Han Dynasty that the number of palaces and gardens hardly increased during the reign of Emperor Wen (180–157 BC). Any project that did not benefit the civilian population was discarded and abolished. Once, the emperor wanted to build a terrace. He summoned the craftsmen and was told the terrace would cost a hundred gold pieces. The emperor replied, "One hundred pieces of gold is the equivalent of the wealth of about ten middle-class families. I have always felt ashamed after inheriting the palace from my father. Why should I even build a terrace?" In his sacred edict, the emperor also demanded that his tomb be built entirely of brick, without any ornamentation of gold, silver, copper, or tin. The tomb was to be built by cutting holes in the mountain, rather than digging piles of earth to build a separate tomb.

In the second year of Zhenguan's reign (AD 628), the weather was hot and humid as summer turned to autumn. A minister wrote to Taizong suggesting that he build a loft to relieve the emperor's rheumatism. Taizong rejected the suggestion, citing the example of Wen's terrace. He said, "Emperor Wen gave up his terrace plan to save the wealth of ten families. How can I put more effort into construction when I am not even close to Wen's virtue? This is simply not the right ideology to take care of the people as an emperor."[18]

Two Song Dynasty books, *Qingxu Miscellaneous Writings Supplement* by Wang Cong and *Words of a Retired Scholar from East of Qi* by Zhou Mi, have recorded incidents of Taizu's frugality. Once, Taizu's secretary told him that the beams in his house needed replacing. No suitable wood could be found at the moment, but there was a large tree that could be felled and used. Song Taizu commented: "Cut the hell out of it! Cut something else!" The practice of cutting large, long pieces into small, short pieces has always been considered a waste of wood and materials.

Frugality, which reflects compassion for people's livelihoods, has long been emphasized as part of the grand construction of every generation of emperors and officials. There are almost endless records of the phrase "humble palace chambers" being mentioned either in the advice of virtuous officials or in the self-effacing statements of emperors.[19] In essence, the goal of merely proposing "humble palace chambers" was never frugality, but the attainment of the highest human morality. It was never about the architecture itself. Rather, it aims to demonstrate the benevolent virtue of self-discipline and frugality that emperors should possess, which is essentially the concept of "benevolence" advocated by Confucianism, a political and moral idea.

Frugality is a constraint on the virtue of emperors. This is not to say that the emperors did not care about the development and extravagance of their palaces. On the contrary, they were particularly active in their ideas of large-scale construction. Emperors have always had a strong sense of luxury. The Epang Palace, the Lishan Mausoleum, the Great Wall, and the Qin Zhi Dao, built by Qin Shi Huang, are all huge projects that required hundreds and thousands of workers. Emperors in later generations were usually frugal and built new developments at the same time, always having an excuse for extravagance.

The *Guanzi* discusses the three ideologies, "Imperial palaces are not built for entertainment but for the well-being of civilians," "High terraces and pavilions, beautiful palace chambers are built to consume wood materials," and "Wood materials are useless if they are not used to decorate palaces." The *I Ching · Xici,*

Part II, says, "Ancient people lived in caves in the wilderness, which later generations replaced with palaces with pillars and roofs to protect them from the wind and rain. This may have been inspired by the symbolism of the ideology of 'Great Strength.'" "Great Strength" suggests that palaces should be built in a grand and solid style to withstand the erosion of wind and rain. They should also have a grand, magnificent, and awe-inspiring style to reflect the majesty and distinction of the ruler, so as to inspire obedience in his civilians.

Around the eighth year of Emperor Gaozu's reign (199 BC), Prime Minister Xiao He mobilized large amounts of manpower and resources to build the Weiyang Palace on top of the Zhangtai Palace. The Weiyang Palace is located to the west of Anmen Avenue in Chang'an City, which is why it was also called the Western Palace. Its total area is about six times that of the Forbidden City. Its overall layout is rectangular, with walls on all sides. The east and west walls are each 2,150 meters long, and the north and south walls are each 2,250 meters long. The entire palace covers an area of about 5 square kilometers, or about one-seventh of the total area of Chang'an. The palace is slightly smaller than Changle Palace, but it has a grander and more magnificent architectural style. When Liu Bang returned from suppressing the rebellion of Xin (the King of Han), he was very angry when he saw the extremely large and extravagant Weiyang Palace that had been built with Xiao He's help. He asked Xiao He, "The world is in turmoil and chaos, and the result of years of war is still unknown. Why would you build such a magnificent and luxurious palace?" Faced with Liu Bang's accusation, Xiao He replied, "The current troubled state of the world is precisely the reason for building the palace now. Besides, the Son of Heaven (Liu Bang) has been traveling all over China. It's necessary to build such a magnificent and luxurious palace to establish the emperor's prestige, which will prevent future generations from surpassing him."[20] Since it's a convincing explanation, Liu Bang accepted it with less reluctance and managed to have both a good reputation and an extravagant palace as emperor.

In fact, emperors have always struggled between the pursuit of "humble palace chambers" and "high terraces and pavilions, beautiful palace chambers," or, in essence, between the contradictory ideas of frugality and extravagance. The emperors wanted to illustrate their benevolence, which included noble morals, self-discipline, frugality, and virtue. However, they could not abstain from luxury either in order to demonstrate their incomparable power and establish their prestige. The traditional Chinese ideology of the "middle way" resolves precisely this contradiction. The middle way, which is the doctrine of the middle, aims to help rulers maintain their frugality without losing their excuse for extravagance and authority. Under the guidance of this ideology, the ancient Chinese rulers finally found the perfect theory that satisfied both sides, which is the appropriate style theory of Chinese architecture.

Mozi, who lived in the same era as Confucius, discussed the problems of economical and suitable palace architecture early on. Mozi said, "Ancient people usually settled in relatively high places or dug caves when they were unable to build houses. The emperors began to build houses because the humidity was harmful to the human body. The construction principle was to build a foundation high enough to avoid dampness, with walls to prevent cold on all sides. The roof should prevent erosion from the weather, and the palace wall should be high enough to separate the interior from the exterior so that people of different sexes could be separated according to ceremonies. These are the requirements that houses should meet."[21] Mozi's view has long been quoted as opposing waste and advocating frugality, while his suggestion of suitability has gone relatively unnoticed. "The roof should prevent erosion from the weather, and the palace wall should be high enough to separate the interior from the exterior so that people of different sexes could be separated according to ceremonies," Mozi suggested the suitability of architecture in terms of functionality, which stands for human need rather than politics.

Dong Zhongshu also argued about the suitability of palace architecture, "A high platform is sunny, with more *yang*, and wide rooms are shady, filled with more *yin*, far from the harmony of heaven and earth. Therefore, the wise do not take extremes, but seek only moderation."[22] Dong Zhongshu explained moderation in terms of harmonizing *yin* and *yang*. He stated that living spaces should be neither too high nor too large and have a reasonable amount of space. His view had a profound influence on Chinese architecture, so much so that even the daily living quarters of the emperors were relatively moderate, with rooms almost the size of ordinary dwellings.

The suitability of functionality was gradually extended to the suitability of rank and status. The emperor's palace and the residences of officials and civilians should all reflect their respective status with different levels of architecture. This effectively prevented people from building large houses and also justified the ruler's need to build palaces that reflected his rule and authority. Until then, a justifiable architectural ideology had been formed between the "humble palace chamber," "high terraces and pavilions, beautiful palace chambers," frugality and extravagance. The ideologies of palace architecture also influenced private architecture in the sense that it should conform to the requirements of rank and status. Such conformity is the "appropriate form."

Chinese ideology and culture have developed mainly through the complementary development of Confucianism, Taoism, and Buddhism. Taoism and Buddhism have historically had some influence on Chinese ideology, but Confucianism remains dominant. It can be concluded that the internal structure of Chinese society is based on Confucianism. The core of Confucianism is the ritual system, which is an unprecedented creation of the Chinese and a special culture in ancient China. "Ritual" is a cultural category with a very broad content. It is not just the manners we often speak of today, nor does it go beyond a state system. Ritual is the core spirit that pervades all Chinese cultural forms, as well as the code of conduct that applies to all areas of ancient social life. In ancient China, the state was governed according to ritual. The state system, the legal system, and all social norms were formulated in accordance with ritual.

In the 11th century BC, the Zhou Dynasty comprehensively summarized the state system, social order, and other ideologies since Xia and Shang and formed a complete national ritual system based on it. It officially became the "ritual" and was used as a form of feudal patriarchal governance. In the Spring and Autumn Period and the Warring States Period, states competed for hegemony to rule the whole country with their own ideology and systems. As a result, there was the "contention of a hundred schools of thought," which gave rise to different schools of ideologies such as Confucianism, Taoism, Legalism, Mohism, and Military Strategists. These schools of thought are also known as the "Hundred Schools of Thought." The situation of ritual governance that had developed in the Zhou Dynasty completely disintegrated, leading to the "collapse of the feudal system and the deterioration of rites and music." The Confucians, led by Confucius, campaigned for the restoration of Zhou rituals and traveled to all states to lobby but were repeatedly frustrated. After Qin unified China, legalism was used to govern the country until Emperor Wu of Han adopted Dong Zhongshu's ideology of "reject the hundred schools, worship only Confucianism." Confucianism then became the orthodoxy of China for more than 2,000 years.

The Confucian classics on ritual include *The Rites of Zhou*, *The Book of Rites*, *The Record of Rituals*, and *The Analects of Confucius*, the first three of which are also known as "The Three Rituals." *The Rites of Zhou* is the uppermost of the three rituals and is also known as *Zhou Officials*. It has six chapters, including "Heavenly Officials: Prime Minister," "Earth Officials: Minister of Education," "Spring Officials: Minister of Worship," "Summer Officials: Minister of War," "Autumn Officials: Minister of Justice," and "Winter Officials: Minister of Public Works," with a total of forty-two volumes. The "Winter Officials" chapter has long been lost and was replaced by "The Record of Examination of Crafts" in the Western Han Dynasty. It is also called "The Record of Examination of Crafts for the Winter Office." Hundreds of workers, supervised by Winter officials, carried out the construction work. *The Rites of Zhou* is a political code for the six officials of Heaven, Earth, Spring, Summer, Autumn, and Winter, that is, the rules and regulations of affairs under the jurisdiction of the six departments. *The Book of Rites* mainly explains the etiquette and relationship guidelines of the scholar-official class during the Spring and Autumn and Warring States periods. It is also known as *The Ritual* or *Rites for the Scholar Class*. *The Record of Rituals* is a collection of essays written by Confucian scholars explaining *The Book of Rites* from the Warring States Period to the Qin and Han dynasties. It is a compilation of materials of Confucian ideologies, which were written by more than one author at different times. The author is not one person, and each person was involved at different times. The majority of the essays may have been written by the seventy-two disciples of Confucius. Other pre-Qin classics have also survived.

The Rites of Zhou and *The Book of Rites* are the institutional composition of the state governance system and ritual. It contains detailed regulations on the social positioning of people and corresponding norms of behavior, covering all aspects of life, such as the construction of capitals and palaces. The houses of princes and learned officials were also bound by the requirements of the state system, while detailed records were developed for the configuration, orientation, and level of all buildings related to people's ritual activities. Ritual is essentially the central idea of Confucianism. It is also the highest guiding ideology[23] and the most important principle of all activities, and its relationship with architecture is no exception.

The ritual system has a direct influence on the layout, pattern, and color of Chinese architecture. Chinese architecture places the main building in the center, with secondary buildings to the left and right in symmetry. Many single-unit buildings form a building complex in an orderly manner. These single-unit buildings are centered around the core building, with a layout of strict directionality, clear hierarchies, and clear levels. The building complex is composed strictly according to the ritual ideology of "there is a proper order to respect superiors and inferiors, a distinction between those above and those below, and a separation between inside and outside." Vertically, the complex sets north as up, east and west as down; horizontally, left is up, right is down; that is, "The north room is the most important, followed by the side rooms, and the seat facing away from the host is for the guest." The location of one's home is a symbol of identity and status. Architectural patterns are also closely linked to the ideology of ritual hierarchy. It imposes strict restrictions on the decoration of the inner and outer eaves, the animal design of the roof tiles, the beam painted decoration, the courtyard decoration, and the interior furnishings. The roof itself is divided into nine levels, the highest of which is the double-eaved hip roof, which can only be used for royal architecture and Confucius Halls. The second level is the single-eave hip roof and the single-eaved gable-and-hip roof, followed by the overhanging and the flush gable roofs. The decorative colors of the buildings are also divided into levels by ritual. In general, yellow is the highest color, followed by red, green, indigo, blue, black, and gray. Palaces usually use gold, yellow, and red, while dwellings can only use black, gray, and white for their walls and roofs.

Sacrifice is also an important expression of the ideology of the ritual system. The ritual was first created during the sacrifice. Therefore, all sacrificial architecture inevitably becomes a part of the important affairs prescribed by the ritual system. Sacrifice originated from the nature and ancestor worship of the ancient people, which focused on the natural earth, mountains, rivers, the state and community, and the ancestors. As a noble ritual system, each type of sacrifice has its own designated building, a special ritual building, which is a unique architectural form in China. "Buildings with high platforms for outdoor sacrifices are called altars (sometimes temples), such as the Temple of Heaven, the Temple of Earth, the Altar of the God of Land and Grain, and the Altar of the God of Agriculture; houses built for worship are called temples, such as the Imperial Ancestral Temple, the Temple of Confucius, the Temple of the God of the Five Great Mountains, and clan temples."[24] Sacrifice is a ritual custom, so sacrificial architecture is not religious. The sacrificial temples are also fundamentally different from the religious temples that developed later. "Sacrificial architecture is a symbol of the state system and the ritual system."[25]

The other important ideologies in the Chinese architectural ideology system are *yin* and *yang*, the Five Elements, and *feng shui*. *Feng shui* is the most specific ideology, concept, or theory most closely related to Chinese architecture. It is impossible to study Chinese architecture without considering it. *Feng shui* is also

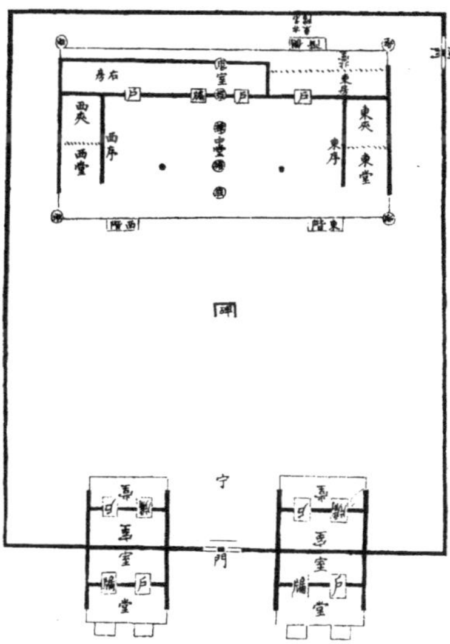

LEFT Layout of a house for learned officials in *The Book of Rituals.*

RIGHT Ancient geomancy: *Imperially Ordered Illustrations of the Book of History.*

known as landscape, geomantic omen, geography, residential divination, yin and yang, azure dragon, etc.

It comes directly from "qi," an ideology of ancient Chinese philosophy. The word first appeared in *The Book of Funerals* by Guo Pu (AD 276–324) of the Eastern Jin period. "When *qi* rides the wind, it disperses; when it reaches the water's edge, it stops. The ancients collected it to prevent its dispersion and directed it to stop, hence the name *feng shui*. In the practice of *feng shui*, the most important thing is to obtain water, and the second is to harness the wind."[26] This is the first time the concept of "qi" is mentioned.

The ancient Chinese divided *qi* into two opposing sides: *yin* and *yang*. The sages generally believed that *qi* exists everywhere. It constitutes all things and is constantly in motion and change. Every condition and matter can be explained by the opposite coexistence and interaction between the *qi* of *yin* and *yang*. Laozi said, "All things carry *yin* and yet embrace *yang*, and by mixing these forces, they achieve harmony."

In fact, the ancient Chinese had already demonstrated a high level of spiritual wisdom before primitive Confucianism and primitive Taoism even existed. They pondered the nature and workings of the world and created three ways of thinking about the universe and all creatures in ancient times, namely the *Yin* and *Yang* Theory, the Five Elements Theory, and the Eight Trigrams Theory. During the Spring and Autumn and Warring States periods, the three theories began to interpenetrate and organically integrate, resulting in the phenomenon of symbiosis of thought, called "The Five Elements of *Yin* and *Yang*" and "The Five Elements of Eight Trigrams."

Yin and *yang* are the two dualities of the world. Everything is created through the combination of *yin* and *yang*, with dualities as two opposites complementing each other.[27] For example, heaven and earth, sun and moon, big and small, long and short, up and down, men and women, etc. The crucial aim of *feng shui* is precisely to deal well with the dualities of *yin* and *yang* through reconciliation. Too much *qi* at either extreme will create an imbalance.

In Chinese architecture, the ideology of *yin* and *yang* and the Five Elements can also be found in various forms, including orientation, mathematics, and symbols.

In terms of spatial orientation, south is *yang*, and north is *yin*; east is *yang*, and west is *yin*. Within the Five Elements, wood, fire, metal, water, and earth correspond to east, south, west, north, and center in the five directions, and indigo, red, white, black, and yellow in the five colors. The ideology of the five directions and five colors then evolved into the "four spirits," namely the Blue Dragon, the White Tiger, the Rosefinch, and the Tortoise, which have become auspicious symbols for determining terrain and orientation in *feng shui*. Each of the "four spirits" represents one direction, with the Blue Dragon in the east, the Rosefinch in the south, the White Tiger in the west, and the Tortoise (whose image is a turtle wrapped by a black snake) in the north. According to the ideology of *feng shui*, the best living environment to choose from is the flowing water of the Blue Dragon in the east, the long road of the White Tiger in the west, the pool of the Rosefinch in the south, and the hill of the Tortoise in the north.[28] Certainly, if the natural environment does not meet one's *feng shui* requirements, the terrain can be altered to suit, for example, by digging lakes or ponds in front of settlements and building mountains behind them. Naming places after their landforms is another way of adapting to *feng shui* symbols. Nanjing Xuanwu Lake (located in the north of Nanjing City during the Ming Dynasty), Xuanwu Gate (located at the north gate of the city), and the Rosefinch Gate (also known as Danfeng Gate or the south gate of the city) are all named under such influence.

Regarding the duality of *yin* and *yang* in numbers, in ancient times, odd numbers were *yang*, and even numbers were *yin*. Nine is considered to be the largest and highest *yang* number, while five is the middle number of all *yang* numbers. Therefore, the combination of nine and five is the most noble and auspicious symbol. That's why emperors are called "The Honour of Nine-Five." Nine has also become the exclusive number for the emperor in architectural design. Designs such as nine-unit width, five-unit depth, nine-step staircase, and nine-row nailed door are all dedicated to imperial buildings. In addition, Chinese buildings tend to have odd-numbered bays. Apart from the need for symmetry around the central axis (the central door of the complex will be off-center when opened if the total number of doors is even), the other main reason is that odd numbers are *yang*.

Based on the fundamental ideology of *yin* and *yang* and the Five Elements, "During the Eastern Han Dynasty (AD 25–220), the practice of divination texts also flourished, combining theories of *yin* and *yang*, *qi* and rhythm as well as witchcraft and prophecy, giving rise to what is known as *feng shui*."[29] Although the three theories of *qi*, *yin*, and *yang*, as well as the Five Elements each have their own origins, they become a unique *feng shui* technique in China when combined. It is also the result of the mixture of objective perceptual experience and subjective idealism, as well as the exchange of simple philosophy and the confusion towards the supernatural. Speaking of the spirit of Chinese architecture, Needham said, "Nowhere are people more enthusiastic than in China in embodying their great vision of the principle that 'man cannot live without nature.' The 'man' he refers to is humanity as a whole. Great buildings such as palaces and temples certainly follow this principle. In urban and rural areas, whether concentrated or scattered, houses often illustrate the sense of 'patterns of the universe' and symbolism through orientations, seasons, wind directions and constellations."[30]

In ancient times, when people's ability to understand science and the world was very low, the "scientific essence" in nature was more difficult for people to discover and understand in relation to superstition. People hoped to fill their inner emptiness and strengthen their belief in survival through superstition. Since ancient times in China, it has been said that "it is better to believe that it exists than to believe that it does not exist." Ideologies such as *yin* and *yang*, the Five Elements, and the Eight Trigrams, with 'qi' as the core and foundation, had a wide and profound impact on Chinese architecture.

In the development process of traditional Chinese cities, whether it's a planned city or a naturally formed village, the selection of sites, the layout, and even the transformation of the environment are all closely related to *feng shui*. The popularity, duration, and impact of the use of *feng shui* has always been unparalleled throughout the development of traditional construction in China and has continued into modern social life. The auspicious and inauspicious perspectives of *feng shui* actually reflect the ancient Chinese view of the environment. For thousands of years, the Chinese have been searching for the ideal living environment. Whether it is a philosophical or secular view of the environment, it is ultimately reflected in the ancient cultural carriers wrapped in *feng shui*, affecting all groups, from emperors and generals to ordinary people.

There is undoubtedly a great deal of controversy surrounding the mysteries and manifestations of *feng shui*. Some people believe that it embodies the simple materialistic philosophy and aesthetics of ancient China. In contrast, others believe that it is completely cultural dross, "Since the Tang Dynasty in the Middle Ages, the concept of *feng shui* has gradually developed and become increasingly complex and absurd, deviating from the orthodox concept of Confucianism. This has led to an increasingly foul atmosphere around *feng shui*, and it has now become a cultural dross embedded in the Chinese people or almost a cultural tumor."[31]

No matter how *feng shui* is viewed today, it is an objective existence. In the thousands of years since its inception, almost all building activities have been carried out under the guidance and influence of *feng shui*. Nowadays, whether it is the study of traditional architecture or the protection of architectural heritage, *feng shui* cannot be ignored. We need to record and analyze the history and the scenes of the past from a historical perspective.

NOTES

1. Zhang Yuhuan, *Question and Answer on Ancient Chinese Architecture* (Beijing: Tsinghua University Press, 2012), 1.
2. Chen Mingda, *Historical Criticism on Ancient Architecture and Sculpture* (Beijing: Cultural Relics Publishing House, 1998), 88.
3. Li Yunhe, *Principles of Classical Chinese Architectural Design* (Tianjin: Tianjin University Press, 2005), 12.
4. Sun Dazhang, *A Brief History of Ancient Chinese Architecture* (Beijing: Tsinghua University Press, 2016), 3.
5. Li Yunhe, *Principles of Classical Chinese Architectural Design* (Tianjin: Tianjin University Press, 2005), 13.
6. Ito Chuta, *History of Chinese Architecture* (Changsha: Hunan University Press, 2014), 3.
7. Joseph Needham, "Civil Engineering and Nautical Technology," in *Science and Civilisation in China*, vol. 4 (Beijing: Science Press, 2008), 126.
8. Liang Sicheng, *The Complete Works of Liang Sicheng*, vol. 4 (Beijing: China Architecture & Building Press, 2001), 14.

9. Liu Zhiping, *Types and Structures of Chinese Architecture* (Beijing: Architectural Engineering Press, 1957), 22.
10. Joseph Needham, "Civil Engineering and Nautical Technology," in *Science and Civilisation in China*, vol. 4 (Beijing: Science Press, 2008), 90.
11. Li Yunhe, *Principles of Classical Chinese Architectural Design* (Tianjin: Tianjin University Press, 2005), 29.
12. Joseph Needham, "Civil Engineering and Nautical Technology," in *Science and Civilisation in China*, vol. 4 (Beijing: Science Press, 2008), 90.
13. Li Yunhe, *Principles of Classical Chinese Architectural Design* (Tianjin: Tianjin University Press, 2005), 30.
14. Ibid., 31.
15. Zhang Jiaji, *Discussions on Chinese Architecture* (Taiyuan: Shanxi People's Publishing House, 2003), 31.
16. Joseph Needham, "Civil Engineering and Nautical Technology," in *Science and Civilisation in China*, vol. 4 (Beijing: Science Press, 2008), 87.
17. Liu Su, *The Civilization of Construction: Traditional Chinese Culture and Traditional Architecture* (Beijing: Tsinghua University Press, 2014), 49.
18. Wu Jing, *Political Maxims of the Zhenguan Era* (Zhengzhou: Zhongzhou Ancient Books Publishing House, 2005), 26.
19. Wang Guixiang, *Ancient Chinese Dwelling Concepts and Architectural Principles* (Beijing: China Architecture & Building Press, 2015), 13.
20. Xu Jialu and An Qiuping, *Full Modern Chinese Translation of 24-Histories* (Beijing: Chinese Dictionary Publishing House, 2004), 138.
21. Mo Di, "Ciguo," in *Mozi*, vol. 1.
22. Dong Zhongshu, "Following the Way of Heaven," in *Chunqiu Fanlu*, vol. 17.
23. Li Yunhe, *Principles of Classical Chinese Architectural Design* (Tianjin: Tianjin University Press, 2005), 42.
24. Liu Su, *Rituals and Architecture* (Beijing: China Architecture & Building Press, 2015), 14.
25. Ibid.
26. Guo Pu, *The Book of Funerals: Outer Chapters*.
27. Liu Su, *The Civilization of Construction: Traditional Chinese Culture and Traditional Architecture* (Beijing: Tsinghua University Press, 2014), 81.
28. Ibid.
29. Wang Guixiang, *Ancient Chinese Dwelling Concepts and Architectural Principles* (Beijing: China Architecture & Building Press, 2015), 177.
30. Li Yunhe, *Principles of Classical Chinese Architectural Design* (Tianjin: Tianjin University Press, 2005), 42.
31. Wang Guixiang, *Ancient Chinese Dwelling Concepts and Architectural Principles* (Beijing: China Architecture & Building Press, 2015), 177.

CHAPTER 2

A HISTORICAL OVERVIEW OF ARCHITECTURE

Chronological Divisions

Although this book is primarily concerned with the study and discussion of ancient Chinese wooden buildings, it is essential to have a basic understanding of the broader context and development of ancient Chinese architecture in order to gain a deeper understanding. Since Mr. Liang Sicheng and Mr. Liu Dunzhen created the discipline of Chinese architectural history in the 1920s, a comprehensive understanding of Chinese architecture has finally embarked on a scientific development. Mr. Liang Sicheng's *The History of Chinese Architecture* and Mr. Liu Dunzhen's *The History of Ancient Chinese Architecture* marked the first comprehensive research into the history of Chinese architecture. Over the following decades, the research on ancient Chinese architecture developed rapidly. However, the research on the division and classification of ancient architecture did not make any major breakthroughs. It was only based on the works of the two pioneers. Mr. Li Yunjing stated, "Chinese architecture has evolved at a gradual pace throughout its history, making it challenging to delineate distinct periods. Some have attempted to categorize this evolution, but their efforts have not always been comprehensive."[1] Although the stages of Chinese architectural history are vaguely defined, it is an important issue that needs to be resolved in order to gain a deeper understanding of the development of Chinese ancient architectural history. In the following section, the key points of historical division are presented for analysis and comparison.

Mr. Liang Sicheng divided the pre-Qing history of Chinese architecture into six periods:[2]

1. Ancient or primitive period (before 200 BC)

Mr. Liang proposed that the literature and tangible objects of this period were both highly lacking, with fewer objects available for investigation before the Yin and Shang dynasties. He also included the Qin Dynasty in this period for its relatively brief history.

2. Han Dynasty (204 BC–AD 220)

The literature that has survived from the Han Dynasty provides a relatively rich record of architecture, though few tangible objects remain. However, a significant number of tomb records, stone carvings, funerary wares, and tombs offer insights into the evolution of previous architectural styles.

3. Wei, Jin, and Southern and Northern Dynasties (AD 220–590)

During the Wei, Jin, and Southern and Northern Dynasties, the development of architecture entered an important period. A significant number of palaces were constructed. In addition, the introduction of Buddhism led to the construction of many Buddhist temples. The existing relics of rock-cut temples and pagodas are all important examples of architecture from this period.

4. Sui and Tang dynasties (AD 590–906)

The Sui and Tang dynasties saw unprecedented progress in the construction of cities and palaces, reaching a peak in the history of Chinese architecture. During this period, there are many important relics, including wooden Buddha halls, stone pagodas, rock-cut temples, and murals, which provide insight into the evolution of Sui and Tang architecture.

5. Five Dynasties and Song, Liao, and Jin dynasties (AD 906–1280)

The architectural style underwent a notable evolution during the Five Dynasties and Song, Liao, and Jin dynasties. It became increasingly refined and delicate, particularly during the Song Dynasty. This period also saw the construction of numerous palaces and prosperous cities. Due to the constant warfare, there are few remaining architectural remains from this period. However, there are numerous examples of towers and timber framing, particularly of wooden structures, which can be visited at archaeological sites. Of greater significance is the fact that the *Yingzao Fashi* was developed during this period, which allows us to gain a deep and comprehensive understanding of ancient buildings.

6. Yuan, Ming, and Qing dynasties (AD 1280–1912)

The Yuan, Ming, and Qing dynasties spanned a significant period of time, during which the general style of buildings remained consistent from dynasty to dynasty. The majority of architectural designs adhered to the original style. During this period, the scale of cities and palace buildings was impressive. In the latter period, some buildings were also influenced by foreign cultures from the West.

Mr. Liu Dunzhen's division of the development of ancient Chinese architecture is roughly the same as that of Mr. Liang in terms of structure:[3]

(1) Three ancient dynasties

(2) Warring States period, Qin and Han dynasties
(3) Jin Dynasty and the Southern and Northern Dynasties
(4) Sui and Tang dynasties
(5) Song, Liao, and Jin dynasties
(6) Yuan, Ming, and Qing dynasties

In his book *An Overview of Ancient Chinese Architecture*, Mr. Fu Xinian outlines the 7,000-year development process of ancient Chinese architectural activities, which has tangible evidence for study, can be divided into five periods, namely the Neolithic Age, Xia, Shang, and Zhou dynasties, Qin and Han dynasties to the Southern and Northern Dynasties, Sui, Tang, and Jin dynasties, and Yuan, Ming, and Qing dynasties. These periods correspond to the infancy, initial formation, basic establishment, flourishing maturity, continuous development, and gradual decline of the ancient Chinese architectural system.[4] Mr. Fu Xinian also suggests that these periods align with the power and unity of the Han, Tang, and Ming dynasties in Chinese history, which have made outstanding achievements in architectural development, especially in scale, technique, and artistic style.

Mr. Sun Dazhang, in his book *A Brief History of Ancient Chinese Architecture*, divides the development of Chinese ancient architecture into five historical periods based on the characteristics and laws of Chinese ancient architecture:[5]

1. Primitive society (600,000 years ago to 21st century BC)

This period marks the infancy of architecture, undergoing a long process of development during which cave and nest dwellings evolved into residential buildings above ground. Houses with different structures emerged in different regions, forming the basis of primitive villages.

2. Slave society (21st century BC to 476 BC, Xia, Shang, Zhou, and Spring and Autumn Period)

Since the Xia Dynasty, China evolved into a class society characterized by private property, hereditary thrones, and extensive use of slave labor. A significant amount of workers were employed to construct buildings such as city walls and palaces. The term "civil engineering" became synonymous with the construction of grand buildings. A particular architectural system emerged for palace construction, with painting and carving already being used to embellish buildings.

3. The early period of feudal society (475 BC–AD 581, Warring States Period, Qin and Han dynasties, Three Kingdoms, Jin, and Southern and Northern Dynasties)

This period saw the emergence of new production relations in Chinese feudal society. The society also underwent its first significant unification and division. By the Han Dynasty, iron had completely replaced bronze as the main production tool. During the Warring States Period, wood-framed buildings were the primary form, while high-platform buildings were widely popular. Images of a variety of architectural styles can be observed in the portraits, murals, and grottoes that survive from the Han Dynasty.

4. The middle period of feudal society (AD 587–1368, Sui and Tang dynasties, Song Liao, Jin, and Yuan dynasties)

During this period, the feudal society underwent a second great unification and the subsequent division. Further adjustments were made to the feudal production relations. The architectural technologies became more sophisticated, and the timber-frame construction system also reached a high level of maturity. During this period many buildings were preserved, which have since become invaluable resources for the study of ancient architecture.

5. The late period of feudal society (AD 1368–1840, Ming and Qing dynasties)

During this period, agriculture and handicrafts reached their highest level of development throughout feudal society, and the final unification of the feudal society was also achieved. This period saw deeper integration and development of a multi-ethnic state. The comprehensive development of architectural technology and art led to remarkable advances in garden art.

In his book *The History of Chinese Architecture*, Mr. Pan Guxi adopted a division method almost the same as that of Mr. Sun Dazhang. The only difference is that the Yuan Dynasty was designated as the late period of feudal society in Mr. Pan's division.[6]

In his book *Chinese Aesthetics*, Mr. Li Yunhe not only expressed his confusion about the division method, but also listed the "views of some other scholars"[7] and then compared them with Western architecture of the same period:

1. Establishment: From the Zhou Dynasty to the Spring and Autumn and Warring States periods (11th century to 3rd century BC), which corresponds to the architectural period of ancient Egypt, West Asia, and Greece.

2. Maturity: The Qin and Han dynasties (3rd century BC–3rd century AD), which corresponds to the Greek and Roman periods.

3. Integration: The period of Wei, Jin, and Southern and Northern Dynasties (3rd to 6th centuries), saw the integration of foreign cultures, which corresponds to the period of early Christian and Byzantine architecture in Europe.

4. Prosperity: The Sui and Tang dynasties (6th to 10th centuries), which corresponds to the period of Byzantine, Roman Nask and early European Gothic architecture.

5. Continuation: The Song, Liao, Jin, and Yuan dynasties (10th century to 14th century), which corresponds to the period of European Gothic architecture.

6. Stagnation: Ming and Qing dynasties (14th century to 19th century), which corresponds to the period of European Renaissance architecture and the various styles that followed.

The different periods of ancient architecture history listed above demonstrate that the general division method follows the development history of ancient Chinese society. This method uses the social form and political process of the early, middle, and late stages of primitive society, slave society, and feudal society as a framework for the division of ancient Chinese architecture. The periods of ancient Chinese architecture are divided according to the natural law of architectural development, which includes infancy, initial formation, basic establishment, prosperous maturity, continuous development, and gradual decline. The two methods of division are both based on the chronicle, which fully describes the development status of ancient Chinese architecture through the process of social, political, economic, and cultural changes and develops a better understanding of the historical development of ancient architecture. However, such division methods also "overlook the difference and asynchrony between the development of architecture and society, as well as the temporal lag of architecture."[8] The division of ancient architectural history determined on the basis of the history of social development cannot fully reveal the inherent development laws and revolutionary highlights of architecture, because the development of architecture and society is not synchronized. Nor can architecture itself, as a tangible entity and culture, be comprehensively summarized by social history.

The continuity and the lack of distinct transformation and style for each era of ancient Chinese architecture compared to Western architecture are precisely the reasons for the difficulty of dividing history according to the logical laws of architectural development. As Mr. Li Yunhe put it, "As for now, it seems to be too early to divide the periods of the history of Chinese architecture. In fact, more comparative analysis should be conducted, and the research requires more historical materials to appear. It is thus difficult to draw practical conclusions in the meantime."[9]

But this does not mean that the issue can be avoided. The division of history should be the first thing to be addressed in order to understand ancient Chinese architecture comprehensively. Therefore, based on the above scholarly division, this chapter will provide a comprehensive review following the clues of the chronicle, while also considering the growth and development logic of ancient architecture. The division will not take into account academic complexities, but acknowledge the growth, maturity, and decline periods of ancient Chinese architecture currently defined by the academia. This approach aims to present a thorough account of the development process of ancient Chinese architecture from ancient times to the present day.

Origins: Prehistoric and Ancient Eras

Primitive society (c. 600,000 years ago to 4,000 years ago)

Primitive society has likely gone through three stages of development: primitive people, matriarchal clan society, and patrilineal clan society. The development of the production tools has gone through two periods: the Paleolithic Period (characterized by chipped stone tools) and the Neolithic Period (characterized by ground stone tools).

In Han Feizi's *Wudu* (*The Five Vermin*), it is mentioned, "In ancient times, the population of people was scarce while birds, beasts, insects, and snakes dominated. A sage emerged and constructed nests out of wood to protect people from the harm caused by these creatures. The people were delighted and appointed him as their ruler, naming him Youchao, meaning 'Master of Nests.'" In *Mozi · Ciguo*, it is stated, "In ancient times, the people lived on hills and mounds, dwelling in caves and holes because they did not yet know how to build palaces and houses. The damp and wet conditions were detrimental to the people's health and well-being. To address this, the sage kings constructed palaces." *The Book of Rites* states, "In ancient times, the early kings did not have palaces. During winter, they would fully utilize caves as their residences, while during summer, they would dwell in nests among the tall trees." From the stories recorded in this ancient literature, we can infer that ancient people may have lived in caves or trees. Archaeological discoveries have corroborated some aspects of these literary works and traditions. The discovery of caves inhabited by Peking Man and primitive people from other areas indicates that natural caves were a relatively common habitat for human beings during this period.

Given that tree nests cannot be preserved as long as natural caves, we believe that early human nests may have existed. The two forms of habitation, cave dwellings (natural caves or caves dug with simple tools imitating animal caves) or nests (including shacks in trees and ground shelters), have also been identified as the two basic forms that first appeared in human architecture.

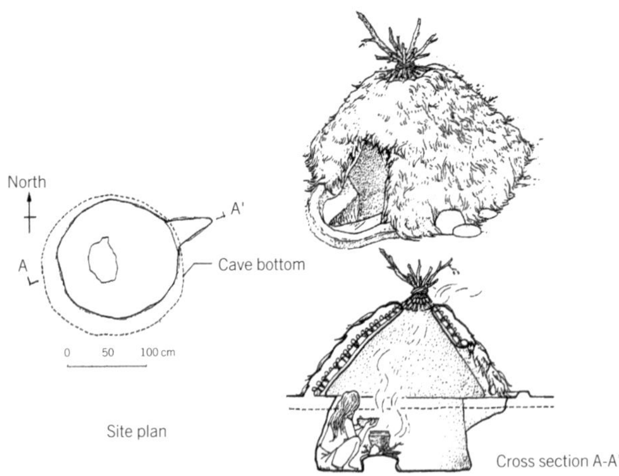

Surmised restoration of semi-cave, Sunqitun site, Jianxi District, Luoyang, Henan Province.

Surmised restoration of primitive houses above ground.

Based on these two fundamental architectural forms, the evolution from nests to *ganlan* architecture gradually took shape in the warm and humid Yangtze River Basin in the south. In contrast, in the middle and lower reaches of the Yellow River in the north, the transition from cave dwellings to structures with wooden frames and mud walls occurred. These transformation processes were characterized by their lengthy and gradual development. The shift from nests to *ganlan* architecture likely began with single-log nests evolving into multi-log nests, followed by stake-supported *ganlan* structures, pillar-supported *ganlan* structures, and ultimately elevated ground-level houses with raised floors, cross-beam, and pillar construction. On the other hand, the evolution of cave dwellings may have progressed from horizontal caves to semi-horizontal caves, vertical caves, semi-caves, vertical half-caves with straight walls, ground-level buildings, and finally compartmentalized structures with wooden frames and mud walls.[10] These hypotheses are based on limited existing illustrative materials.

Until about six or seven thousand years ago, most of China had evolved into a clan society. The illustration of the result of this transformation has just recently been discovered through the continuous discovery of sites and relics; that is to say, we have found the results of the late evolution in the related sites of the late Neolithic period of primitive society.

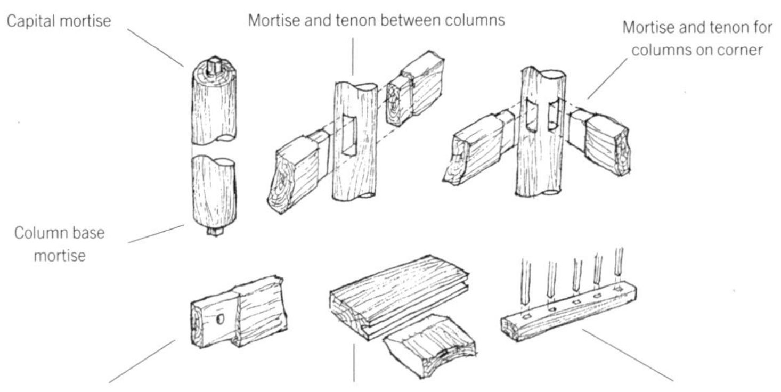

Mortise and tenon in wood components, the Hemudu site, Yuyao, Zhejiang Province.

Hemudu culture, which existed about 7,000 to 5,000 years ago (5000–3300 BC). A typical architectural site of the culture is the Hemudu site in Yuyao, Zhejiang, which is the earliest example of the use of the mortise and tenon technique to build timber-framed houses. Components of wooden architecture such as pillars, beams, and mortise-and-tenon planks have been excavated. It is speculated to be an elongated, large-scale *ganlan*-style building.

Yangshao culture, which existed about 7,000–5,000 years ago (5000–3000 BC). Typical architectural sites of the Yangshao culture include the Jiangzhai site and the Banpo site.

The Jiangzhai Yangshao Cultural Village site in Lintong, Shaanxi Province, which existed between 4600 BC and 4400 BC, presents a residential area of five house groups. Each group sets a large house as its core, surrounded by smaller houses, reflecting the living conditions of the clan community.

The Banpo Yangshao Cultural Village site in Xi'an, Shaanxi Province, which existed between 4700 BC and 4000 BC, has a residential area that is surrounded by artificial ditches. The area is divided into two districts by a small moat. Each district has a large house in the center surrounded by small studios. The houses are of two types: semi-sunken and ground-level buildings.

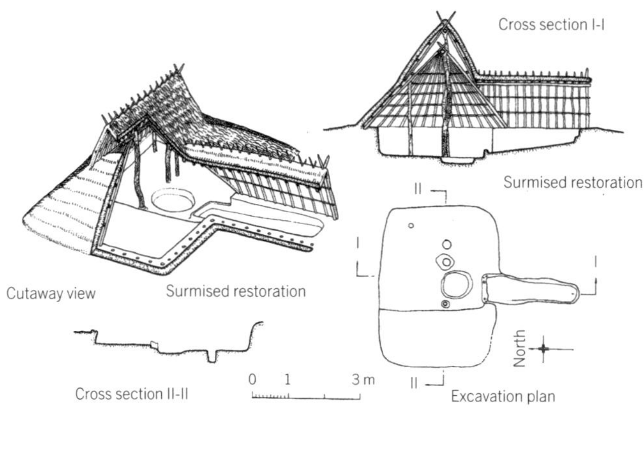

Surmised restoration of primitive square houses, Banpo site, Xi'an.

In terms of construction technology, the buildings of the later stage of Yangshao culture have progressed from semi-sunken dwellings to ground-level buildings, with houses divided into several rooms. The houses are of rectangular and circular layouts, with timber frames and mud walls. The interior walls are hardened and leveled by smearing them with fine mud or by grilling the floor. In the late stage, the columns are in a neat array with clear functions designated for

Excavated Chengziya Neolithic site, Zhangqiu District, Jinan, Shandong Province.

Yuchi Temple site, Mengcheng County, Anhui Province.

the wooden frame and the exterior wall. The construction technology of wooden architecture reached an unprecedented level.[11]

The Dawenkou culture existed about 6,500–4,500 years ago (4500–2500 BC). During this period, not many architectural ruins were discovered. Some typical sites include the mid-to-late-stage Chengzi site in Zhucheng City, Shandong Province, the Dadunzi site in Pizhou City, Jiangsu Province, and the Yuchi Temple site in Mengcheng, Anhui Province.

The Longshan culture existed approximately 4,500–4,000 years ago (2500–2000 BC). The main architectural ruins of this period include the Keshengzhuang site in Xi'an, the Hougang site in Anyang, Henan, and the Taosi site in Xiangfen, Shanxi Province.

A suite-style semi-cave dwelling with two connected rooms was found at the Keshengzhuang site in Xi'an. It has a layout in the shape of the Chinese character "Lü" (吕) and differs the inner room from the outer room.

At the Hougang site in Anyang, a ground-level building was found. The foundation of the house is made of rammed earth and the walls are made of timber frame and mud walls. The interior wall is whitewashed with foundation stone under the columns.

Patterns engraved on the whitewash walls were discovered at the Taosi site in Xiangfen, Shanxi Province, which are believed to be the oldest interior decoration known to date.

During the same period, a circular stone hut was found in northern China, which is believed to be a sacrificial house.

Surmised restoration of villages at the Yuchi Temple site, Mengcheng County, Anhui Province.

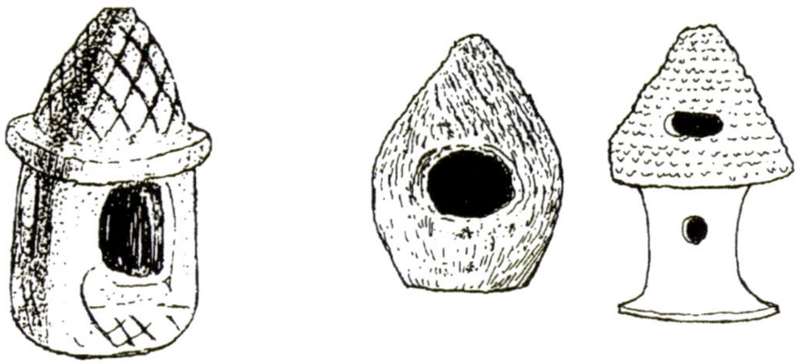

TOP Surmised restoration of primitive villages.

BOTTOM Models of Neolithic pottery houses unearthed in China.

Xia, Shang, and Zhou dynasties (c. 21st century–771 BC)

Xia Dynasty (about 2070–1600 BC). In the 21st century BC, the Xia Dynasty was established. Since then, China has turned into a slave society, where property is privatized and thrones are inherited. Although the historical facts of the Xia Dynasty are recorded in ancient literature, the existence of the Xia has always been highly controversial in archaeological discoveries. The Yanshi Er'litou site in Luoyang City, Henan Province, is considered to be the key to resolving the issue of Xia culture. It was here that the earliest palace city in China was first discovered, with its palace building complex and workshop areas laid out along the central axis. Palace No.1 of the complex has an opening as wide as eight rooms, each of which is relatively uniform. The columns are neatly lined up and surrounded by corridors, with doors facing south.

Shang Dynasty (c. 1600–1046 BC). The Shang Dynasty entered the mature stage of slave society, during which stone tools were replaced by bronze. The discovery of oracle bone inscriptions has provided us with a clear written record. The capitals of the Shang Dynasty were all cities built with high walls, large palace complexes, and additional gardens and terraces, using rammed earth construction techniques of the time. The relatively representative architectural sites of the Shang Dynasty include the Shangcheng site in Zhengzhou, the Shangcheng site in Shigou County, Yanshi, Henan (located 5,000 meters east of the Erlitou site. Some experts believe that it is the capital city of Bo, which was built after Shang destroyed the Xia Dynasty), Xiaotun Yinxu in Anyang, Henan,

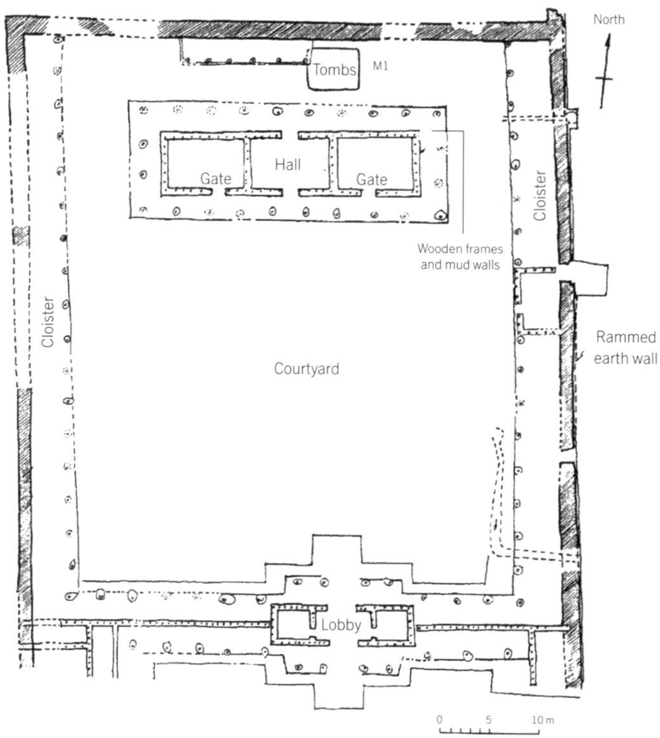

Plan of the Excavation of the Palace of the Xia Dynasty at the Erlitou site, Yanshi, Luoyang, Henan Province.

and the Panlong Shangcheng site in Huangpi County (now Huangpi District), Hubei Province. Among them, the main palace with courtyard buildings found at the Shangcheng site in Shigou County, Yanshi, is 90 meters long and is the largest Early Shang Dynasty building site known to date.

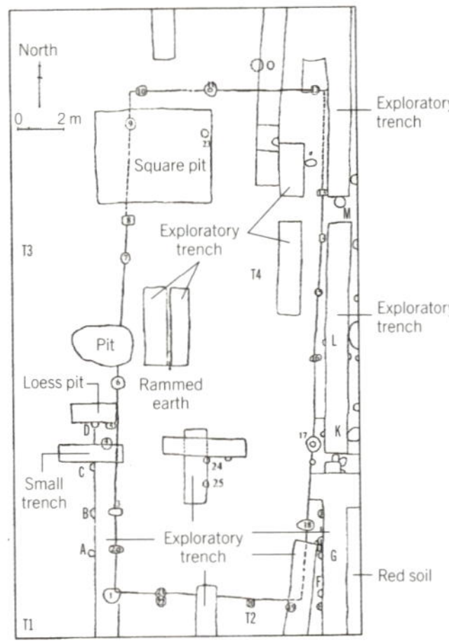

LEFT Excavated site of the palaces of Yin Dynasty, Anyang, Henan Province, in the spring of 1929 (leftmost Li Ji).

RIGHT Plan of the excavation site of the Shang Palace of the Yin Dynasty, Anyang, Henan Province.

LEFT A surmised model of the Shang Palace of the Yin Dynasty, Anyang, Henan Province.

RIGHT An imitated building modeled after the Shang Palace of Yin Dynasty, Anyang, Henan Province.

Western Zhou Dynasty (1046–771 BC). Zhou destroyed Shang and established a feudal system. Many vassal states of royal families and nobles were established in various regions, with more construction activities than in the previous dynasty. The capital of the Western Zhou Dynasty is Feng and Hao, located on both sides of the Fenghe River in the southwest of Xi'an. Although the layout of Feng and Hao has not been fully verified in archaeology, it has detailed written records. As recorded in *The Rites of Zhou · Kao Gong Ji*, "Craftsmen build the city, square with nine li on each side, with three gates on each side; within the city, nine vertical and nine horizontal streets, and the roads are nine tracks wide; the ancestral temple on the left, the altar of the soil and grain on the right, facing the court in front and the market behind; the market and court are governed by one person." This is also the earliest record of the layout of Chinese cities.

Representative architectural sites of the Western Zhou Dynasty include the early Zhou Dynasty ruins of Fengchu Village in Qishan, Shaanxi, the *ganlan*-style wooden architectural sites in Qichun, Hubei, and the Zhaochen site in Fufeng, Shaanxi. In the Western Zhou Dynasty, the production and use of ceramic began. It was also found that brick pavement, rammed earth walls, and adobe walls with a three-layered plaster made of white lime, sand, and yellow clay, whose surface was flat and level, were used in the same period.[12] The structural framework based on wooden beams and the courtyard layout had

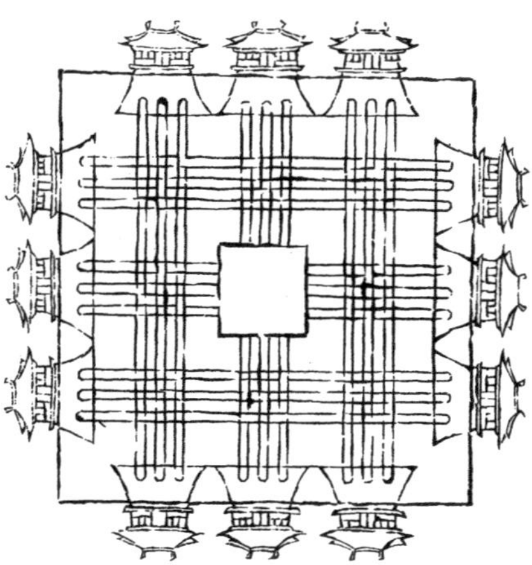

Layout of the city of Emperor Zhou in *Pictures of Three Rituals*.

already been formed in the Western Zhou Dynasty.[13] The images of brackets set between columns and *dou* on top of columns appeared on the bronze artifacts of the Western Zhou Dynasty, which is also believed to be the earliest known form of *dougong*.

Ling-gui

Square caldron with beast's feet

Architectural components represented on Western Zhou Bronze Wares.

Spring and Autumn and Warring States periods (770–221 BC)

Spring and Autumn (770–476 BC). The royal family of Zhou gradually declined, and feudal production relations began to appear with the use of iron tools and cattle. It is said that Luban lived in this period when the capital city was divided into two cities, the big city and the small city. The big city was for residence, and the small town was for palaces. The "li" (里), a centrally enclosed residence based on the grid, and the "shi" (市), which operated at regular hours, had both appeared. A "curfew" was also implemented at night.

Representative architectural sites of the Spring and Autumn Period include the ancient capital of Jin in Houma, Shanxi Province, the ancient city of Eastern Zhou in Luoyang, Henan, Yong City of Qin in Fengxiang, Shaanxi, and the ancient capital of Ying of Chu in Jiangling, Hubei. Tiles, cylindrical tiles, and slab tiles were used in all of the above sites. Particularly, cyan-gray, blue-gray hollow bricks with hard surfaces, and decorative patterns were also found in the Yong City of Qin in Fengxiang; rammed earth platforms were discovered at the ancient capital of Jin in Houma. Elevated palaces for political, military, and recreational purposes (multi-story buildings built in a stepped rammed earth pattern, also called terraces) have already appeared. In *The Analects of Confucius*, it writes, "Zang Wenzhong, a senior official in the state of Lu, built a divination house for the big tortoise (for divination) from Cai. The house features a bracket set carved in the shape of a mountain and beams and columns painted with algae. How can such a person be considered wise?" In *Guliang Commentary on the Spring and Autumn Annals*, it states, "In autumn, they painted the palace columns red. According to the ritual, the columns of the emperor and feudal lords should be black, those of high-ranking officials should be gray, and those of common soldiers should be dark-colored. Red columns are not in accordance with the proper rituals," "Carving the beams in a certain way is not in accordance with the proper rituals." It can be seen that both architectural decoration and color developed to a certain extent during this period.

Warring States Period (475–221 BC). During this period, agriculture, handicrafts, and trade developed more; the widespread use of iron tools promoted the development of architectural technology. The palace adopted molded floor tiles with patterns. The floor and stairs were made of hollow bricks with vermilion painted on the surface. The plain walls were painted with frescoes. Many large capitals appeared during this period, especially those of the Seven Powerful States of the Warring States Period: Linzi of Qi (Linzi, Shandong), Handan of Zhao (Handan, Hebei), Yanying of Chu (Jingzhou, Hubei), Daliang of Wei (Kaifeng, Henan), Ji of Yan (Fangshan, Beijing), Xinzheng of Han (Xinzheng, Henan), and Xianyang of Qin (Xianyang, Shaanxi).

At present, the ruins that have been confirmed to be representative architecture from Warring States cities include the ruins of the ancient capital of Qi in Linzi, Shandong, the ruins of the ancient capital of Zhao in Handan, Hebei, the ruins of the lower capital of Yan in Yixian, Hebei, the ruins of Xianyang Palace of Qin State in Xianyang, Shaanxi, and the ruins of the ancient city of Zhongshan in Pingshan, Hebei. At the site of the ancient capital of Zhao in Handan, there are more than ten elevated buildings, indicating that elevated buildings were still very popular during the Warring States Period, and they were equipped with heating and drainage systems. At the site of the ancient city of Zhongshan, a copper plate with a cemetery drawing and measurements was found. It seems to be an architectural planning drawing and also indicates that construction was carried out according to the drawing at that time.

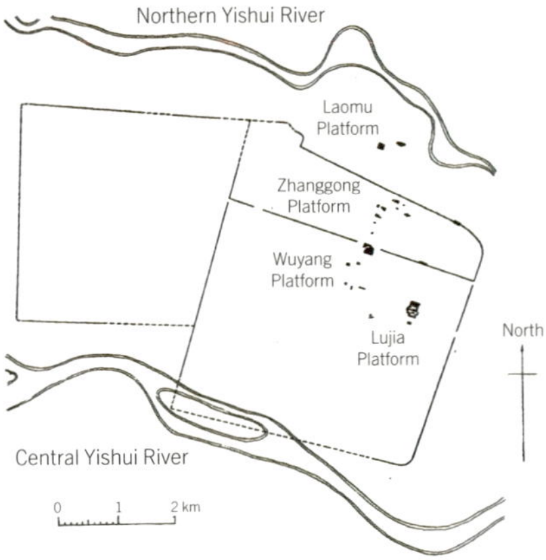

TOP The site and its measured drawing of the lower capital of Yan, Yi County, Hebei Province.

MIDDLE & BOTTOM The site of the ancient capital of Zhao, Handan, Hebei Province.

Development: Qin, Han, and Three Kingdoms

Qin (221–206 BC). Although the Qin Dynasty didn't last for long, it gained remarkable national power after unifying the six states. It built roads to reach the whole country, the Great Wall to defend against the Huns, and a new palace on the south bank of the Weishui River in Xianyang that copied the palaces of the six states. The scale of construction was unprecedented, and it also advanced the communication, integration, and development of building construction in various areas.

The major architectural sites of Qin include the famous Epang Palace, the Mausoleum of the First Qin Emperor, and the Great Wall of the Qin Dynasty, etc. The ruins of Epang Palace ruins and Mausoleum of the First Qin Emperor are still under excavation. The rammed earth platform remained in the Afang Palace is of great scale and is speculated to be an unfinished palace. In the Mausoleum of the First Qin Emperor, a large number of decorative bricks, tiles, and stones have been unearthed, as well as bronze lintels adorned with cloud patterns, stone-carved drainage channels, and other architectural components.

Han (206 BC–AD 220). The reign of the Han Dynasty lasted for more than 400 years. It was also the second most powerful feudal dynasty after Qin. The Han Dynasty can be divided into two periods: the Western Han Dynasty and the Eastern Han Dynasty. The Western Han Dynasty was founded by the Great Ancestor Liu Bang and lasted for 13 generations, a total of 214 years (206 BC–AD 8); the Eastern Han Dynasty was founded by Liu Xiu of the Han imperial

Tiles in the Qin Dynasty.

family. He restored Han rule by overthrowing the 15-year-long new dynasty established by Wang Mang to replace the Western Han Dynasty. The Eastern Han Dynasty lasted for 13 generations with a total of 196 years (AD 25–220).

The architectural technology and aesthetics of the Han Dynasty have reached a very high level, which can be attributed to the long-term architectural practices of numerous types and quantities at that time.[14] Although there are no surviving wooden buildings from the Han Dynasty, a great deal of information can be gathered from the excavation of tombs, portraits on stone bricks, murals, and architectural sites from that period.

The four basic architectural forms of beam and column, interlocking bucket, *ganlan*, and *jinggan* already existed in the Han Dynasty. The structure of wooden beams and columns was widely used at that time. Roof shapes included single-sloped roofs, double-sloped suspended roofs, cornered roofs, pyramidal roofs,

Watchtower, burial object in Han tombs, Gaotang County, Shandong Province

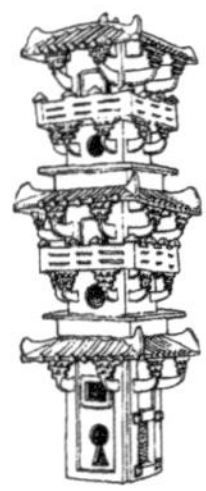

Watchtower, burial object in Han tombs, Wangdu County, Hebei Province

Watchtower, funerary ware in Han tombs, Shanzhou District, Sanmenxia, Henan Province

Que, portrait brick, Chengdu, Sichuan Province

Houses inside *wubao*

Wubao (fortified manor house), burial object in Han tombs, Guangzhou, Guangdong Province

A collection of buildings, engraved stones, Suining County, Jiangsu Province

Images of Han architecture in engraved stones and burial objects.

and four-cornered roofs, among which four-cornered roofs were predominantly used for higher buildings.[15] Through the study of the remaining pottery houses and stone *que* (a type of ceremonial gateway), it can be learned that the four-cornered roof was used because of the common square layout of buildings at that time, where the diagonal lines intersected at 45 degrees, making it unsuitable for the gable-and-hip roof form. The roofs were constructed using a two-layer overlapping technique. The ridge decorations of surviving Han Dynasty pottery buildings closely resemble those found later in Tang Dynasty buildings. In the Han Dynasty, the number of bays of rooms varied between odd and even numbers. Most palaces and residential buildings had odd-numbered bays, while the tombs and sacrificial buildings mostly adopted even-numbered bays.

There is clear evidence of the use of *dougong* in the Han Dynasty. From the early practice of stacking horizontal beams on top of vertical beams in *ludou* (capital block) structures, to the transformation of stacked beams into spaced rectangular blocks, the small rectangular blocks evolved into the early form of *xiaodou* (interlocking bracket). In comparison to later bracket forms, *ludou* and *xiaodou* lack the *doushen* (bracket body) and *douyi* (bracket head). Bracket sets often took the form of one *dou* with two ascending pieces or one *dou* with three ascending pieces, the latter being more common in later periods. There is no evidence of bracket sets projecting outward, and they were mostly used as under-eave bracket sets and flat-seat bracket sets. During the Han Dynasty, the structure of wooden columns and beams had become quite mature. The columns were relatively uniform in shape, and there were column bases elevated above the ground. Double columns were used at the corners. In Ban Gu's *Ode to the Western Capital*, there is mention of "arching the rainbow beam against the dragon," referring to a curved beam. Although there is no concrete evidence available, this indicates that besides straight beams, curved beams were also used for decorative purposes during the Han Dynasty.

The most important architectural ruin of Han cities is Chang'an City of Han. The site of Chang'an City of Han was built in the first year of the reign of Liu Ying, Emperor Hui (194 BC). However, its main palaces, Changle Palace and Weiyang Palace, were originally built during the reign of Liu Bang, Emperor Gaozu of Han. The construction method of "palaces first, capitals later" was relatively rare in ancient times.[16] There are three gates on each side of Chang'an

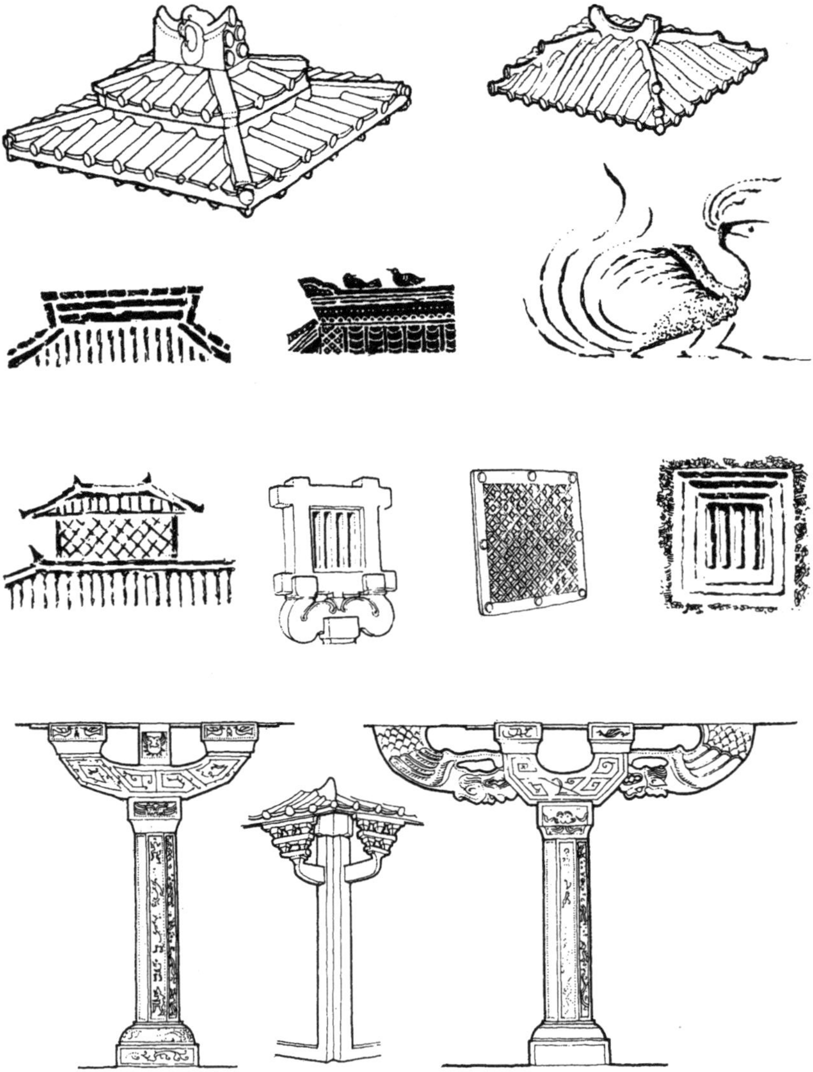

Octagonal column, ancient stone relief tombs, Yinan County, Shandong Province

Square twin columns, burial object, Wangdu County, Hebei Province

Octagonal column, ancient stone relief tombs, Yinan County, Shandong Province

Details of Han architecture.

City in the Han Dynasty. Inside the city, there are also Mingguang Palace, Gui Palace, and North Palace, with Jianzhang Palace built outside the western city wall of the city, adjacent to the Imperial Shanglin Garden. Sacred buildings such as Mingtang, Biyong, and the Imperial Ancestral Temple were built in the southern suburbs of Chang'an. Mingtang and Biyong were built according to the "round sky and square earth" pattern, which is the earliest discovery of this type of architecture. Luoyang City was called Dongdu during the Western Han Dynasty and was the auxiliary capital of Chang'an City; in the Eastern Han Dynasty, Luoyang City replaced Chang'an City and became the new capital. Its palace area is located in the northwest of the city, where the South Palace and the North Palace are located. Another relatively prominent architectural site of Han cities is the ruins of Yecheng. Yecheng was the fiefdom of Cao Cao in the late Han Dynasty. There was an avenue running through the east and west of the city, dividing it into north and south districts. In the center of the northern district is the palace, while the southern district consists of neat blocks. Government and judicial offices were located along the palace avenue. This type of urban structure with centralized functions had a certain influence on the later development of the city.

The residential buildings of the Han Dynasty also had no remains. Their architectural images can thus only be inferred from the information reflected in the ruins, literature, portraits, stone bricks, tombs, and stone *que*s. Among the relatively centralized and representative Han village sites is the Han Dynasty village site of Sanlizhuang, Neihuang County. The site remains relatively complete, potentially because it was in ruins after being flooded by the Yellow River. The tombs of the Western Han Dynasty emperors were mainly located in the hilly areas on the north bank of Chang'an and Weishui and southeast of Chang'an City. In addition, the Eternal Mausoleum of the Retired Emperor is located in Yueyang (30 kilometers north of Lintong District, Shaanxi Province). The tombs of the Eastern Han Dynasty are mainly located in the southeast and northwest suburbs of Luoyang, with the Mausoleum of Liu Xie, the Last Emperor of Han, located in Zhuolu City, which is 150 kilometers north of Luoyang. The tombs of princes and lords consist of Beidong Mountain in Xuzhou, Han Tombs in Guishan, Baiji, Shizi Mountain, Yinan Han Tombs, and Mancheng Han Tombs in Hebei Province. The remains of Han *que* are mostly distributed in Sichuan Province, among which the most prominent are Gaoyi Tomb *Que* in Ya'an, Feng Huan *Que* in Qu County, Lord Shen Tomb *Que*, Lord Pingyang *Que* in Mianyang, Li Ye *Que* in Zitong, etc. In Chongqing Province, there are Ding Fang *Que* and Wuming *Que* in Zhongxian County. In Shandong Province, there are Jiaxiang Wu Family *Que* in Jining City and Huangshengqing *Que* and Gongcao *Que* in Pingyi County. Taishi *Que*, Shaoshi *Que*, and Qimu *Que* are located in Dengfeng County (now Dengfeng City) in Henan Province. Currently, only one ruin of tomb architecture, an above-ground stone temple, remains—the Guo Family Ancestral Hall in Xiaotang Mountain, Shandong Province. This is the only evidence that Han Dynasty architecture was of relatively high quality.

Although Buddhism spread to Luoyang in the late Eastern Han Dynasty, there are almost no examples of Buddhist architecture and related architecture left. Buddhist architecture was described in the relevant literature as follows: in the tenth year of Emperor Ming of the Eastern Han Dynasty (AD 67), Tianzhu monks She Moteng and Zhu Falan built the White Horse Temple in front of

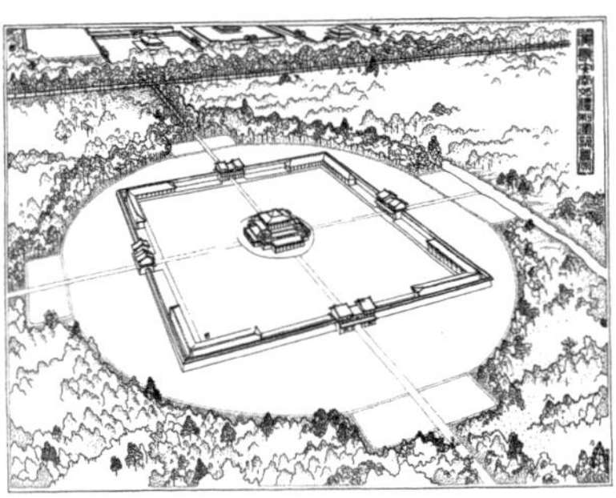

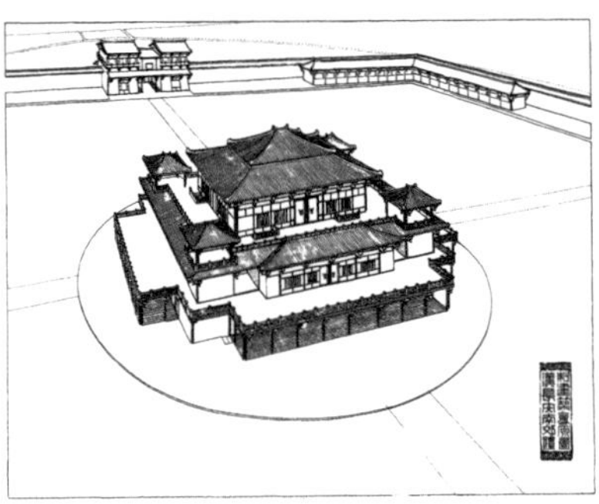

Restored images of Chang'an City in the Han Dynasty.

Stone Shrine of the Guo Family in Xiaotang Mountain, Shandong Province.

Remains of a Han Dynasty village, Sanlizhuang, Neihuang County, Hebei Province.

Remains of a Han Dynasty well, Sanlizhuang, Neihuang County, Hebei Province.

Patterns of tiles in the Han Dynasty.

the Yong Gate of Luoyang (the existing White Horse Temple was rebuilt after destruction); in the final years of the Eastern Han Dynasty, a pagoda shrine was built in Xiapi, Xuzhou by Ze Rong. In addition, a Buddhist stone carving remains in Lianyungang, which is the cliff sculpture on Kongwang Mountain.

Three Kingdoms (AD 220–280). At the end of the Eastern Han Dynasty, the Yellow Turban Rebellion broke out, pitting powerful leaders from all regions against each other. China entered a turbulent period of separatism and division among warlords. In AD 220, Cao Pi, the son of Cao Cao, replaced the Han Dynasty with the Wei Kingdom as the new emperor. In AD 221, Liu Bei became emperor and founded the Shu Kingdom. In AD 222, Sun Quan established himself as the emperor of the Wu Kingdom, and the confrontation between the three states began. The 61 years between the founding of Wei by Cao Pi in AD 220 and the destruction of Wu by Western Jin in 280 are known as the "Three Kingdoms Period." As a powerful political force, the three regimes had already begun building their capitals before AD 220. In AD 196, Cao Cao unified North China and built the city of Xuchang. He gained control of the city of Yecheng and made it the new capital in AD 204. In AD 208, Sun Quan built the city of Jianye (present-day Nanjing). All of these can be regarded as the architecture of the Three Kingdoms, which not only marks the decline of the Han Dynasty architecture, but also serves as the beginning of the architecture of later dynasties.

The architecture of the Three Kingdoms is poorly preserved. From historical literature, it can be seen that the major cities built by each state include Yecheng City and Xuchang City of Cao Wei, Luoyang City of Wei-Jin, Chengdu City of Shu Han, and Jianye City of Sun Wu. In recent years, more relics from Yecheng and Luoyang City have been excavated. Yecheng is located northwest of Anyang, Henan Province. In the ninth year of Emperor Xian's Jian'an era (AD 204), Cao Cao gained control of Yecheng and started construction in its original form. When Cao Pi established Luoyang as his capital in AD 220, Yecheng was designated as an auxiliary capital. Cao Wei carried out construction and development in Yecheng for 17 years in a row. Most of the ruins of Yecheng were buried in the silt, with no other ruins above ground except for two platforms, Jinfeng and Tongjue, and the remains of Taiwu Hall.[17] The existing ruins were excavated in the 1980s. There was an avenue connecting the east and west of the city, dividing it into the north and south districts. The center of the northern district was the palace, and the southern district consisted mostly of neat blocks. Government and judicial offices were located along the avenue leading to the palace. To show that Wei was the successor to the Han Dynasty, Cao Pi established Luoyang as the state capital in AD 220. It was rebuilt from the ruins of Luoyang, which had been burned down after Dong Zhuo forced Emperor Xian to move the capital to Chang'an in AD 190. The new capital existed for 91 years until it was burned down again in AD 311 during the Western Jin Dynasty. The city walls and gates of Luoyang, rebuilt by Cao Wei, were based on the ruins of the Eastern Han Dynasty. The southern palace of the Eastern Han Dynasty was abandoned, and the northern palace was restored, forming an overall layout with the palace in the north, the official office and the residential area in the south, and the road leading south from the gate of the main hall of the northern palace as the central axis.

The architecture of the Three Kingdoms mainly continued the old architectural system in the Eastern Han Dynasty. Due to the large-scale destruction caused by the war, it is possible that some architectural techniques were lost while the others were innovated through reconstruction. Although there are Han *fu* (rhapsodies) such as "Wei Capital Rhapsody," "Jingfu Hall Rhapsody," "Wu Capital Rhapsody," and "Shu Capital Rhapsody," which describe the architecture of the capital cities during the Han Dynasty, the scenes of this period can only be speculated upon because there are no physical remains of the architecture. In the context of Han *que*, there are structures equivalent to the Ya'an Gaoyi *Que* and the Pingyang Lord Pingyang *Que* of the Three Kingdoms period. Gao Yi, the prefect of Yizhou during the Han Dynasty, died one year after the Battle of Red Cliffs (AD 209), and Li Fu (Lord Pingyang), the prefect of Baxi, died in the first or second year of Yanxi of Shu Han (AD 238 or 239). The two *que*s built for Gao Yi and Li Fu can also be considered as architectural relics of the Three Kingdoms period. The architectural style reflected in these two *que*s shows little difference from the *que* structures of the Eastern Han Dynasty.

The brick and stone structures of the Han Dynasty were mainly used for tombs. However, from the excerpts in the *Tai Ping Yu Lan* (*Imperial Readings of the Taiping Era*) that quote descriptions of the brick ground structures of

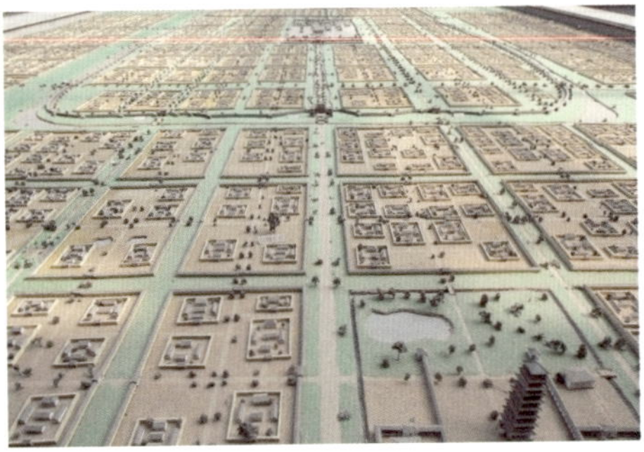

LEFT Surmised city model of Yecheng of the Cao Wei Dynasty.

RIGHT Remains of Jinfeng High Platform, one of the three platforms in Yecheng of the Cao Wei Dynasty.

the Lingyuntai of Cao Wei's Luoyang Palace from the *Shu Zheng Ji*, it can be learned that during this period, the development of brick structures and their ground application provided a technical basis for the brick pagodas of the Jin Dynasty and later.[18] There are not many records of Buddhist buildings of the Three Kingdoms in the literature. We only know that Sun Quan, Lord of Wu, built the Jianchu Temple and later the Ashoka Pagoda in Jianye, which is the earliest record of a Buddhist pagoda in the region south of the Yangtze River.[19]

Integration: Two Jin and Southern and Northern Dynasties

In AD 265, Sima Yan established the Jin Dynasty to replace Wei by abdication. In AD 280, he destroyed Wu, unified the entire country, and enfeoffed the kings of feudal states after destroying Wu. Less than ten years later, the "Rebellion of the Eight Kings" broke out as all the kings vied for power. The peoples of northern China took the opportunity to take over the Central Plains and successively established a series of separatist regimes known in history as the "Sixteen Barbarian States." In AD 316, the Western Jin Dynasty fell, and the remaining forces of the country established a state south of the Yangtze River. This state was called the Eastern Jin Dynasty, with Jiankang as its capital. The Eastern Jin

Dynasty was succeeded by Liu Song in AD 420. Later, after the reigns of the Song, Qi, Liang, and Chen dynasties, it became the Southern Dynasty when Sui unified the entire country in AD 589, which lasted for 169 years. In AD 439, the Northern Wei regime of the Xianbei people gradually unified northern China.[20] Later, due to internal divisions, the regime split into Eastern Wei and Western Wei, which were eventually destroyed by Northern Qi, and after the Sui Dynasty unified the country in AD 589, Northern Zhou became the Northern Dynasty.

The Two Jin and the Southern and Northern Dynasties were the most chaotic periods in Chinese history, lasting more than 300 years. During this time, wars were frequent, and social and economic development was severely affected and damaged. In comparison, the northern Central Plains suffered the most. Due to the large number of people who migrated south from the Central Plains, there were relatively few wars and less chaos in the south, leading to better economic development than in the north. The most notable event during this period was the flourishing of Buddhism, followed by the integration of other foreign cultures. In addition, the introduction of northern minority cultures had a significant impact on Chinese architecture. The prolonged war also contributed to the prevalence of the study and refined intellectual discussion of Daoism among scholar-officials. The ideals of purity, loftiness, and natural simplicity to some extent dominated the intellectual landscape, and exerted a significant influence on various aspects of Chinese architectural forms, including gardens. From the

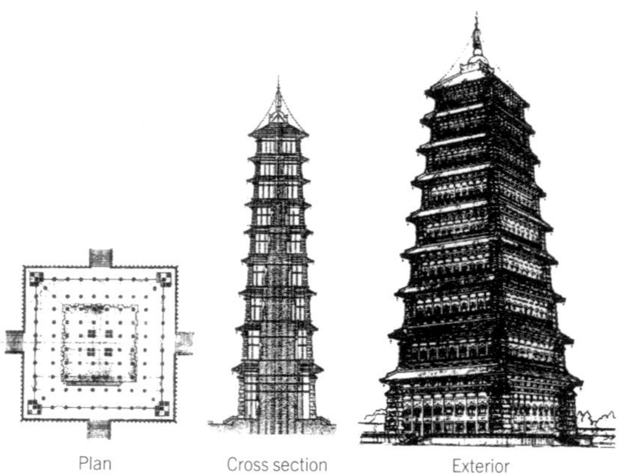

Yongning Temple Pagoda of Northern Wei, Luoyang (Reconstruction drawing by Yang Hongxun).

perspective of the overall development of Chinese architecture, this period marks an important transition. The architectural style of the Han and Wei dynasties gradually faded, while the architecture of the Sui and Tang dynasties began to flourish. This marks the beginning of a glorious era in Chinese architecture.

During the Two Jin and Southern and Northern Dynasties, Buddhist architecture flourished, and a large number of temples, pagodas, and grottoes were built throughout the country.

Literature indicates that a significant number of multi-story wooden pavilion-style towers were constructed. It can also be inferred that the construction technique of twisted columns was adopted, and the construction technique of brick structures has reached a very high level. The Songyue Temple Pagoda is the oldest surviving Buddhist pagoda in China. Built in the fourth year of Zhengguang, Emperor Xiaoming of the Northern Wei Dynasty (AD 523), this pagoda is dodecagonal with stepped eaves and retains traces of foreign architectural influence. It plays an important role in the history of Chinese architecture.

The construction of temples and pagodas reached its peak during the Northern Wei Dynasty in Luoyang and the Southern Dynasties in Jiankang. According to the *Record of Buddhist Monasteries in Luoyang*, there were a total of 1,367 temples in Luoyang. The renowned Tang Dynasty poet Du Mu wrote in his famous poem "Spring in the South of the Yangtze River": "Four hundred and eighty temples of the Southern Dynasties, amid fog and rain, how many towers and pavilions." The *History of the Southern Dynasties* also records that during the reign of Emperor Wu of the Liang Dynasty, there were over five hundred temples in the capital city. It is evident that "four hundred and eighty temples in the Southern Dynasties" is not merely a literary embellishment. The most remarkable temple among them is the Yongning Temple Pagoda built in the first year of the Xiping era of Emperor Xiaoming's reign (AD 516). This pagoda is described in detail in the *Record of Buddhist Monasteries in Luoyang*, "It was constructed with wooden frames, reaching a height of ninety *zhang*, with an additional ten *zhang* with the finial, making it a total of one thousand *chi* from the ground. Located one hundred *li* away from the capital, it is visible from a distance … On top of the finial, there is also a golden treasure vase." The text also describes, "Amidst high winds and through the endless night, the precious bells resonate with a clear and melodious sound that can be heard over a distance of more than ten *li*."* The Yongning Temple Pagoda, which would have stood approximately 300 meters tall by today's standards, was a testament to the advanced wooden construction technology of its time. Unfortunately, it was destroyed by a massive fire in the third year of Yongxi era of Emperor Xiaowu's reign (AD 534). "The fire lasted for three months and even burned into the ground, resulting in smoke lingering for a year." The pagoda no longer exists today.

Grotto temples that were of great importance in the Two Jin and Southern and Northern Dynasties include the Kizil Caves which were first built at the end of the 3rd century, the Mogao Caves in Dunhuang which were first built in the second year of the founding of the former Qin Dynasty (AD 366), and the Yungang Grottoes that were first built in the Heping era of Emperor Wencheng of the Northern Wei Dynasty (AD 460–465). The Longmen Grottoes were first built after Emperor Xiaowen of the Northern Wei Dynasty moved his capital to Luoyang (AD 493). The Tianlongshan Grottoes were first built in the Eastern Wei Dynasty (AD 534–550), and the Xiangtangshan Grottoes were first built during the reign of Emperor Wenxuan of the Northern Qi Dynasty (AD 550–559). In

TOP Cave No. 16, Tianlongshan Grottoes, Taiyuan, Shanxi Province.

BOTTOM Part of the southern wall of Guyang Cave, Longmen Grottoes, Henan Province.

* In ancient Chinese measurements, 1 *zhang* is approximately equal to 3.33 meters, 1 *li* is approximately 500 meters, 1 *chi* is approximately equal to 33.3 centimeters, and 1 *cun* is one-tenth of a *chi*, or approximately 3.33 centimeters.

addition, there are also other grottoes with fewer architectural features, such as the Maijishan Grottoes, which were built in the Later Qin Dynasty (AD 384–417), and the Tiantishan Grottoes, which were first built in Northern Liang (AD 397–460), the Bingling Temple Grottoes that was first built in the first year of Jianhong (AD 420) in the Western Qin Dynasty, and the Gongxian Caves that was first built in the second year of Xiping in the Northern Wei Dynasty (AD 517). Representative grottoes in southern China include the Qianfo (Thousand Buddha) Cliff on Qixia Mountain in Nanjing. This grotto was built during the second year of Yongming (AD 484) by Emperor Wu of Qi. However, it has lost its original appearance due to severe damage.

On top of that, the existing small stone pagoda of Chongfu Temple in Shuozhou, built in the second year of the Tian'an era of Xianwen Emperor in the Northern Wei Dynasty (AD 467),[21] Yicihui Stone Column built in the fifth year of Tiantong in the Northern Qi Dynasty (AD 569), and the wood effect sarcophagus in the Luoyang Ning Mao Stone Tomb built in the second year of Yong'an era of the Xiaozhuang Emperor in the Northern Wei Dynasty (AD 529),[22] as well as the stone animals and Shinto pillars scattered in the tombs of the emperors of the Southern Dynasty in Nanjing, Danyang, and other places, are all solid examples for the study of Jin and Northern Dynasty architecture.

During the Two Jin and Southern and Northern Dynasties, fully adorned decorative bricks were commonly used in the tombs, while the portrait bricks

LEFT Tombstone of Xiao Jing, Nanjing, Southern Dynasties.

MIDDLE Yicihui Stone Column, Hebei Province, North Qi Dynasty.

RIGHT Stone Pagoda, Chongfu Temple, Shuozhou, Shanxi Province, Northern Wei Dynasty.

used in the Han Dynasty have completely disappeared. In the Great Mausoleum of the Southern Dynasties of Xishan Bridge of Nanjing Museum, the *Seven Sages of Bamboo Grove Painting*, which is made of molded bricks, still exists. With the rise of Buddhism and the construction of grotto temples, the stone carving technology developed significantly compared to previous dynasties. Great changes took place in the architectural form. The extensive use of *xumizuo* base (Sumera seat), flame-style arches, and bundled lotus-style columns in architecture, along with decorative elements such as passion fruit, lotus, and scroll grass, enriched the themes of architectural decoration and had a significant influence on the subsequent development of architecture.

Flourishing: Sui, Tang, and Five Dynasties

In AD 581, Emperor Wen Yang Jian ended the Northern Zhou Dynasty and established the Sui Dynasty. In AD 589, he traveled to the south to conquer the state of Chen and unified the entire country. Later, the Sui Dynasty was replaced by the Tang Dynasty, founded by Li Yuan in AD 618, which lasted for 37 years. The Tang Dynasty reigned for 288 years before yielding to the Later Liang in AD 906. China entered a tumultuous period marked by division and frequent changes of dynasties until Zhao Kuangyin established the Song Dynasty in AD 960, a reign that endured for 54 years. This epoch is famously known as the "Five Dynasties and Ten Kingdoms" period.

The Sui and Tang dynasties represent the golden age of Chinese feudal society and the mature period of the development of ancient Chinese architecture. During this period, a complete Chinese architectural system was formed and reached an unprecedented level by absorbing the influence of foreign architecture and culture based on the architecture of previous dynasties. The architecture of the Tang Dynasty also had a significant impact on the architecture of Korea, Japan, and other regions.

The important sites of capitals of the Sui and Tang dynasties include the Daxing City (Chang'an City of Tang) of Sui and Dongdu City of Sui (Luoyang City of Tang), as well as Jiangdu City (now Yangzhou) of Sui. Major palace sites include Daxing Palace (Taiji Palace of Tang) of Sui, Dongdu Palace (Luoyang

Palace of Tang), Daming Palace of Tang, and Xingqing Palace of Tang, etc. The capitals of the Sui and Tang dynasties were the largest cities in the world at that time.

There are relatively more pagoda ruins from the Sui and Tang dynasties. They were built in brick, stone, and wood and designed in single-story, pavilion, and multi-eave styles, including:

(1) The Four-Door Pagoda of Shentong Temple in Jinan,[23] Shandong, a pavilion-style stone pagoda built in the seventh year of the Daye era of the Sui Dynasty (AD 611)
(2) The Dushun Pagoda in Xi'an, Shaanxi Province, a pavilion-style brick pagoda built in the nineteenth year of the Zhenguan era of the Tang Dynasty (AD 645)
(3) The Xuanzang Pagoda[24] of Xingjiao Temple, Xi'an, Shaanxi, a pavilion-style brick pagoda built in the second year of the Zongzhang era under the reign of Emperor Gaozong of the Tang Dynasty (AD 669)
(4) The Great Wild Goose Pagoda[25] of Da Ci'en Temple in Xi'an, Shaanxi Province, a pavilion-style brick pagoda built during the Chang'an era of Wu Zetian (AD 701–704)
(5) The Small Wild Goose Pagoda of the Jianfu Temple in Xi'an, Shaanxi Province, a multi-eave brick pagoda built in the first year of the Jinglong era of Emperor Zhongzong of the Tang Dynasty (AD 707)
(6) The Fawang Pagoda of Xianyou Temple in Zhouzhi County, Xi'an, Shaanxi Province, a pavilion-style brick pagoda built in the thirteenth year of the Kaiyuan era of the Tang Dynasty (AD 725)
(7) The Jingzang Zen Master Stupa of Huishan Temple in Dengfeng, Henan Province, a pavilion-style brick pagoda built in the fifth year of the Tianbao era of Emperor Xuanzong of the Tang Dynasty (AD 746)
(8) The Huichong Pagoda of Lingyan Temple in Changqing District, Jinan, Shandong, a pavilion-style brick pagoda built during the Tianbao era of the Tang Dynasty (AD 742–756)
(9) The Nine-Pinnacle Pagoda[26] in Licheng District, Jinan, Shandong, a multi-eave brick pagoda built in the fifteenth year of the Tianbao era of Emperor Xuanzong of the Tang Dynasty (AD 756)

(10) The Xiuding Temple Pagoda[27] in Anyang, Henan, a pavilion-style brick pagoda built in the first year of the Qianyuan era of Emperor Suzong of the Tang Dynasty (AD 758)
(11) The Lamp-Burning Pagoda[28] of Faxing Temple in Zhangzi Shanxi Province, a stone pagoda with an eternal flame built in the eighth year of the Dali era of the Tang Dynasty (AD 773)
(12) The Fanzhou Zen Master Pagoda[29] in Yuncheng, Shanxi Province, a pavilion-style brick pagoda built in the second year of the Changqing era of Emperor Muzong of the Tang Dynasty (AD 822)
(13) The Qianxun Pagoda of Chongsheng Temple in Dali, Yunnan Province, a multi-eave brick pagoda built during the reign of King Quan Fengyou of Nanzhao (AD 824–859)
(14) The Minghui Master Pagoda in Pingshun, Shanxi Province, a pavilion-style brick pagoda built in the fourth year of the Qianfu era of Emperor Xizong of the Tang Dynasty (AD 877)
(15) The Iron Tower of Guangxiao Temple[30] in Guangzhou, Guangdong, a pavilion-style iron tower built in the sixth year of the Dabao era of the Southern Han Dynasty (AD 963)
(16) The Sheli Pagoda of Qixia Temple in Nanjing, Jiangsu, a multi-eave stone pagoda built in the Southern Tang Dynasty of the Five Dynasties (AD 937–975)
(17) The Fawang Temple Pagoda in Dengfeng, Henan, a multi-eave brick pagoda believed to be from the Tang Dynasty

In addition, the Foguang Temple Shanxi Province on Mount Wutai from the eleventh year of the Dazhong era of the Tang Dynasty (AD 857) and its Dharani pillar, built in the fourth year of the Qianfu era of the Tang Dynasty (AD 877), hold a prestigious position in Chinese architectural history.

Ancient Chinese architecture developed and established its system during the Sui and Tang dynasties. This was a significant and transformative period during which wooden buildings left their first physical relics. With the aid of relevant literature, we were able to confirm the actual locations of ancient architecture and verify previous speculations.

TOP LEFT The Four-Door Pagoda of Shentong Temple, Shandong Province, Sui Dynasty.

BOTTOM LEFT The Nine-Pinnacle Pagoda, Licheng District, Jinan, Shandong Province, Tang Dynasty.

TOP RIGHT Dharani pillar of Foguang Temple, Wutai Mountain, Shanxi Province, Tang Dynasty.

BOTTOM RIGHT Lin Huiyin surveying this Dharani pillar.

At present, there are seven known surviving wooden buildings from the Tang Dynasty:

(1) The Main Hall of Nanchen Temple in Wutai Mountain, Shanxi Province, built in the third year of the Jianzhong era of the Tang Dynasty (AD 782)
(2) The Main Hall of the Five Dragons Temple in Ruicheng, Shanxi Province, built in the sixth year of the Dahe era of the Tang Dynasty (AD 832)
(3) The Scripture Library of Qinglian Temple in Zezhou County, Shanxi Province, built in the seventh year of the Dahe era of the Tang Dynasty (AD 833)
(4) The East Hall of Foguang Temple in Wutai Mountain, Shanxi Province, built in the eleventh year of the Dazhong era of the Tang Dynasty (AD 857)
(5) The eaves of Cave 196 of the Mogao Caves in Dunhuang, Gansu Province, built in the second year of the Jingfu era of the Tang Dynasty (AD 893)

East Hall, Foguang Temple, Wutai Mountain, Shanxi Province.

TOP LEFT Inspection of the Main Hall of Nanchan Temple, Wutai Mountain, Shanxi Province.

TOP RIGHT Scripture Library of Qinglian Temple, Zezhou County, Shanxi Province.

BOTTOM LEFT Front Hall of Yuhuang Temple, Bu Village, Zhangzi County, Shanxi Province.

BOTTOM RIGHT Bell Tower of Kaiyuan Temple, Zhengding County, Hebei Province.

(6) The Front Hall of the Yuhuang Temple in Bu Village, Zhangzi County, Shanxi Province, inferred to date back to the Tang Dynasty
(7) The Bell Tower of Kaiyuan Temple in Zhengding County, Hebei Province, inferred to date back to the Tang Dynasty[31]

At present, there are five known remains of wooden buildings from the Five Dynasties:

(1) The West Side Hall of Longmen Temple in Pingshun, Shanxi Province, built in the third year of the Tongguang era of the Later Tang Dynasty (AD 925)

(2) The Main Hall of Tiantai Temple in Pingshun, Shanxi Province, dated to the fourth year of the Changxing era of the Later Tang Dynasty (AD 933)[32]

(3) The Amitābha Hall of Dayun Temple in Pingshun, Shanxi Province, built in the third year of the Tianfu era of the Later Jin Dynasty (AD 938)

(4) The Wanfo Hall of Zhenguo Temple in Pingyao, Shanxi Province, built in the seventh year of the Tianhui era of the Northern Han Dynasty (AD 963)

(5) The Dacheng Hall of the Confucian Temple in Zhengding County, Hebei Province, inferred to date back to the Five Dynasties (AD 907–960)[33]

West Side Hall of Longmen Temple, Pingshun County, Shanxi Province.

TOP Main Hall of Tiantai Temple, Pingshun County, Shanxi Province.

BOTTOM Dacheng Hall of Confucian Temple, Zhengding County, Hebei Province.

Amitābha Hall of Dayun Temple, Pingshun County, Shanxi Province.

The ancient wooden buildings of the Sui and Tang dynasties have simple structures without redundant decorative components. Building components have reached a considerable level of standardization with a clearer hierarchical system of buildings. The columns and beams used in the construction are all very sturdy, especially the imposing *dougong* at the chapiter. Structures of moderate importance, such as palaces, governmental offices, and temples, incorporate *dougong* elements. During the Sui and Tang dynasties, *dougong* brackets took on the form of inverted-V. The columns have side brackets, and the walls are built in the *shoufen* style.* Therefore, the building has a solid and stable appearance.[34] The elevation of the columns gives a certain curvature to

* *Shoufen* (收分) refers to the architectural technique of tapering or reducing the dimensions of structural elements to enhance both aesthetic appeal and structural stability, commonly seen in traditional Chinese architecture.

the eaves, with the central ridges rising at both ends. In this way, the buildings don't appear to be cumbersome despite their considerable volume. The roofs of the Sui and Tang dynasties continued to use flat tiles and pan tiles as in the previous dynasties. Glazed tiles were used in palaces and temples, especially in later periods. Plank doors and straight lattice windows were still sometimes used in decoration, with wooden structures brushed with vermilion and walls brushed with white. Sometimes brackets were brushed in khaki to enhance the sense of layers. From the late Southern and Northern Dynasties to the golden age of the Tang Dynasty, curves had an increasing presence in architectural appearance and architectural components, which led to the development of two methods: "scroll-ending" and "lift-and-fold." Both methods standardized the curvature in architecture and brought about certain changes in architectural styles as a whole.

Continuation: Liao, Song, and Jin Dynasties

The Song, Liao, and Jin dynasties lasted for 300 years, during which China experienced a historic period of confrontation between multi-ethnic regimes. The Song Dynasty (Northern Song [AD 960–1127], Southern Song [AD 1127–1279]) of the Han regime, Liao (AD 916–1125) of the Khitai regime, Jin (AD 1115–1234) of the Jurchen regime, and Xixia (AD 1038–1227) of the Tangut regime coexisted successively. As a state of multiple regimes, the Song Dynasty has always been perceived as a weak state that humiliatingly surrendered to the Yuan. However, with the development of the previous dynasties, the Song Dynasty actually reached the highest stage of development of Chinese feudal society in terms of culture, economy, science and technology. Mr. Chen Yinke once said, "After thousands of years of development, Chinese culture reached its peak in the Zhao and Song dynasties." It was also in the Song Dynasty that Chinese architecture made significant achievements.

Bianliang, the former regional capital of the Northern Song Dynasty, was established as the state capital, also known as Bianjing (present-day Kaifeng City, Henan Province; *jing* means capital in Chinese). Paintings depicting Bianjing, such as *Along the River during the Qingming Festival* and *The Eastern Capital: A*

Dream of Splendor by Meng Yuanlao, are extremely useful materials for studying the urban architecture of Bianjing. The Northern Song Dynasty established Luoyang as Xijing (Western Capital), Yingtian Fu as Nanjing (Southern Capital), and Daming Fu as Beijing (Northern Capital). After the House of Zhao (the imperial clan of Song) moved south, Lin'an was built as the new capital. Liao and Jin each established five capitals.

There are relatively more sites of ancient wooden buildings from the Song, Liao and Jin dynasties, which have provided us with a large number of research examples. Since most of the remains of ancient wooden buildings have undergone repairs several times in the course of history, the degree of complexity and uncertainty in dating them has also increased. In addition to the cases where the time of construction is clearly recorded in inscriptions and related materials, the time of construction is often inferred from the characteristics of the wooden structure. In order to better understand and study the architecture of this period, an incomplete list of wooden buildings from the period is compiled as follows, among which there are conflicting opinions about the date of construction of some buildings.

The existing major sites of wooden buildings from the Liao and Song dynasties:

(1) The Main Hall of Hualin Temple in Fuzhou, Fujian, built in the second year of the Qiande period of the Northern Song Dynasty (AD 964)[35]

(2) The eaves of caves 427 and 437 of Mogao Caves in Dunhuang, Gansu Province, built in the third year of the Kaibao period of the Northern Song Dynasty (AD 970)

(3) The Middle Hall of Chongming Temple in Gaoping, Shanxi Province, built in the fourth year of the Kaibao era of the Northern Song Dynasty (AD 971)

(4) The Sleeping Hall of Jidu Temple in Jiyuan, Henan Province, built in the sixth year of the Kaibao era of the Northern Song Dynasty (AD 973)

(5) The eaves of Cave 444 of Mogao Caves in Dunhuang, Gansu Province, built in the ninth year of Kaibao in the Northern Song Dynasty (AD 976)

(6) The Front and Middle Halls of North Jixiang Temple in Lingchuan, Shanxi Province, built in the third year of Taiping Xingguo in the Northern Song Dynasty (AD 978)
(7) The eaves of Cave 431 in the Mogao Caves in Dunhuang, Gansu Province, built in the fifth year of Taiping Xingguo in the Northern Song Dynasty (AD 980)
(8) The Guanyin Pavilion of Dule Temple in Ji County, Tianjin, built in the second year of the Tonghe era of the Liao Dynasty (AD 984)
(9) The Pilu Hall of Youxian Temple in Gaoping, Shanxi Province, built in the Chunhua era (AD 990–994)
(10) The Main Hall of Mei'an Temple in Zhaoqing, Guangzhou, Guangdong, built in the second year of the Zhidao era of the Northern Song Dynasty (AD 996)
(11) The Maitreya Buddha Pagoda of the Laojun Temple in Dunhuang, Gansu Province, built in the third year of the Xianping era of the Northern Song Dynasty (AD 1000)
(12) The Scripture Library of Anchan Temple in Taigu, Shanxi Province, built in the fourth year of the Xianping era of the Northern Song Dynasty (AD 1001)
(13) The Main Hall of Baoguo Temple in Ningbo, Zhejiang Province, built in the sixth year of the Dazhong Xiangfu era of the Northern Song Dynasty (AD 1013)
(14) The Sanqing Hall of Yuanmiao Taoist Temple in Putian, Fujian, built in the eighth year of the Dazhong Xiangfu era in the Northern Song Dynasty (AD 1015)
(15) The Qianfo Hall of the Chongqing Temple in Changzi, Shanxi Province, built in the ninth year of the Dazhong Xiangfu era of the Northern Song Dynasty (AD 1016)
(16) The Main Hall of the City God Temple in Ruicheng, Shanxi Province, built in the Dazhong Xiangfu era of the Northern Song Dynasty (AD 1016 the latest)
(17) The Main Hall of Fengguo Temple in Yi County, Liaoning Province, built in the ninth year of the Kaitai era of the Liao Dynasty (AD 1020)

(18) The Main Hall of Jiwang Temple in Wanrong, Shanxi Province, built in the first year of the Tiansheng era of the Northern Song Dynasty (AD 1023)

(19) The Shengmu Hall of Jinci Temple in Taiyuan, Shanxi Province, built in the Tiansheng era of the Northern Song Dynasty (AD 1023–1032)

(20) The Great Hall of Xingfan Temple in Qi County, Shanxi Province, built in the third year of the Tiansheng era of the Northern Song Dynasty (AD 1025)

(21) The Passage Hall of South Jixiang Temple in Lingchuan, Shanxi Province, built in the eighth year of the Tiansheng era of the Northern Song Dynasty (AD 1030)

(22) The Bojiajiaocang Hall of Huayan Temple in Datong, Shanxi Province, built in the seventh year of the Chongxi era of the Liao Dynasty (AD 1038)

(23) The Main Hall of Shousheng Temple in Xiangning, Shanxi Province, built in the first year of the Huangyou era of the Northern Song Dynasty (AD 1049)

(24) The Mani Hall, Main Gates, and Zhuanlunzang Pavilion of Longxing Temple in Zhengding, Hebei, built in the fourth year of the Huangyou era of the Northern Song Dynasty (AD 1052)

(25) The Sakyamuni Pagoda of Fogong Temple in Ying County, Shanxi Province, built in the second year of the Qingning era of the Liao Dynasty (AD 1056)

(26) The Great Hall of Kaihua Temple in Gaoping, Shanxi Province, built in the sixth year of the Xining era of the Northern Song Dynasty (AD 1073)

(27) The Jade Emperor Hall of the Jade Emperor Temple Hall in Fucheng, Zezhou, Shanxi Province, built in the ninth year of the Xining era in the Northern Song Dynasty (AD 1076)

(28) The Yuanjue Hall of Faxing Temple in Changzi, Shanxi Province, built in the fourth year of the Yuanfeng era in the Northern Song Dynasty (AD 1081)

(29) The Worship Pavilion and Main Hall of Dongyue Temple in Zhou Village, Zezhou, Shanxi Province, dated to the fifth year of the Yuanfeng era of the Northern Song Dynasty (AD 1082)
(30) The Main Hall of Jingde Temple, Gaodu, Zezhou, Shanxi Province, estimated to date back to the second year of the Yuanyou era in the Northern Song Dynasty (AD 1087)
(31) The Shakya Hall of Qinglian Temple, Zezhou, Shanxi Province, built in the fourth year of the Yuanyou era in the Northern Song Dynasty (AD 1089)
(32) The Corner Hall of Jindong Temple in Xinzhou, Shanxi Province, built in the eighth year of the Yuanyou era of the Northern Song Dynasty (AD 1093)
(33) The Main Hall of Er'xian Temple in Jincheng, Shanxi Province, built in the fourth year of the Shaosheng era of the Northern Song Dynasty (AD 1097)
(34) The Great Hall of Longmen Temple in Pingshun, Shanxi Province, built in the fifth year of the Shaosheng era of the Northern Song Dynasty (AD 1098)
(35) The Guanyin Pavilion and Dizang Pavilion of Qinglian Temple in Zezhou, Shanxi Province, built in the first year of the Yasukuni era of the Northern Song Dynasty (AD 1101)
(36) The Jade Emperor Hall of the Jade Emperor Temple in Beiyi City, Zezhou, Shanxi Province, built in the fourth year of the Daguan era of the Northern Song Dynasty (AD 1110)
(37) The Jidu Temple in Xidun Village, Zezhou, Shanxi Province, built in the fourth year of the Xuanhe era of the Northern Song Dynasty (AD 1122)
(38) The Main Hall of Emperor Guan Temple in Yangquan, Shanxi Province, built in the fourth year of the Xuanhe era of the Northern Song Dynasty (AD 1122)
(39) The Wuliang Hall of Emperor Guan Temple in Dingxiang, Shanxi Province, built in the fifth year of the Xuanhe era of the Northern Song Dynasty (AD 1123)

(40) The Main Hall of Chuzu Temple in Dengfeng, Henan, built in the seventh year of the Xuanhe era of the Northern Song Dynasty (AD 1125)
(41) The Dedication Hall and Sleep Quarter of Sima Qian Temple in Hancheng, Shaanxi Province, built in the first year of the Jingkang era of the Northern Song Dynasty (AD 1126)
(42) The Main Hall of Emperor Guan Temple in Guangrao, Shandong, built in the second year of the Jianyan era of the Southern Song Dynasty (AD 1128)
(43) The Bailin Temple in Wudu, Gansu, built in the ninth year of the Qiandao era of the Southern Song Dynasty (AD 1173)
(44) The Sanqing Hall of Xuanmiao Temple in Suzhou, Jiangsu, built in the sixth year of the Chunxi era of the Southern Song Dynasty (AD 1179)
(45) The Feitianzang Hall of Yunyan Temple in Jiangyou, Sichuan, built in the eighth year of the Chunxi era of the Southern Song Dynasty (AD 1181)
(46) The Wenshu Hall of Geyuan Temple in Laiyuan, Hebei Province, inferred to date back to the Liao Dynasty
(47) The Great Hall of Shanhua Temple in Datong, Shanxi Province, inferred to date back to the Liao Dynasty
(48) The Main Hall of Kaishan Temple in Gaobeidian, Hebei, inferred to date back to the Liao Dynasty
(49) The Main Hall of Chongjiao Temple in Guyi Village, Changzhi, Shanxi Province, inferred to date back to the Song Dynasty
(50) The Main Hall of Puguang Temple in Baidao, Shouyang, Shanxi Province, inferred to date back to the Song Dynasty
(51) The Front Hall of Jade Emperor Temple in Shangdang District, Changzhi, Shanxi Province, inferred to date back to the Song Dynasty
(52) The Shakya Hall of Chongshou Temple in Zezhou, Jincheng, Shanxi Province, inferred to date back to the Song Dynasty
(53) The Main Hall of Er'xian Temple in Xiaonan Village, Zezhou, Jincheng, Shanxi Province, inferred to date back to the Song Dynasty

(54) The Main Hall of Chengtang Temple, Shuanghedi Village, Dadonggou, Zezhou, Jincheng, Shanxi Province, inferred to date back to the Song Dynasty

(55) The Main Hall of the Earth God Temple in Nanshe Village, Gaodu, Zezhou, Jincheng, Shanxi Province, inferred to date back to the Song Dynasty

(56) The Main Hall of Yuanqi Temple in Lucheng, Shanxi Province, inferred to date back to the Song Dynasty

(57) The Pilu Hall of Zisheng Temple in Gaoping, Jincheng, Shanxi Province, inferred to date back to the Song Dynasty

(58) Dai Temple in Zezhou, Jincheng, Shanxi Province, inferred to date back to the Song Dynasty

(59) The Main Hall of Zhaoren Temple in Changwu, Shaanxi Province, inferred to date back to the Song Dynasty

(60) The Main Hall of Yuqing Buddhist Temple in Xia County, Shanxi Province, inferred to date back to the Song Dynasty

Mani Hall of Longxing Temple, Zhengding County, Hebei Province.

TOP Library Building of Longxing Temple, Zhengding County, Hebei Province.

BOTTOM Main Gate of Longxing Temple, Zhengding County, Hebei Province.

LEFT Shengmu Hall of Jinci Temple, Taiyuan, Shanxi Province.

FOLLOWING PAGE Shengmu Hall of Jinci Temple, Taiyuan, Shanxi Province.

TOP Qianfo Hall of Chongqing Temple, Zhangzi County, Shanxi Province.

BOTTOM Main Hall of Kaihua Temple, Gaoping, Shanxi Province.

Main Hall of Baoguo Temple, Ningbo, Zhejiang Province.

(61) The Main Hall of Er'xian Temple in Xiaohuiling Village, Lingchuan, Jincheng, Shanxi Province, inferred to date back to the Song Dynasty
(62) The Main Hall of Biyun Temple, Xiaozhang Village, Changzi, Shanxi Province, inferred to date back to the Song Dynasty
(63) The Main Hall of Xuancheng Temple in Xijianzi Village, Yuci, Shanxi Province, inferred to date back to the Song Dynasty

The wooden buildings of the Song Dynasty that are located in Jiangnan (the region south of the Yangtze River):[36]

(1) The Main Hall of Hualin Temple in Fuzhou, Fujian, built in the second year of the Qiande era of the Northern Song Dynasty (AD 964)
(2) The Main Hall of Mei'an Temple in Zhaoqing, Guangzhou, Guangdong, built in the second year of the Zhidao era of the Northern Song Dynasty (AD 996)
(3) The Main Hall of Baoguo Temple in Ningbo, Zhejiang Province, built in the sixth year of the Dazhong Xiangfu era of the Northern Song Dynasty (AD 1013)
(4) The Sanqing Hall of Yuanmiao Temple in Putian, Fujian, built in the eighth year of the Dazhong Xiangfu era of the Northern Song Dynasty (AD 1015)
(5) The Great Hall of Shisi Temple in Jingning, Lishui, Zhejiang, built in the tenth year of the Shaoxing era of the Southern Song Dynasty (AD 1140)
(6) The Sanqing Hall of Xuanmiao Temple in Suzhou, Jiangsu, built in the sixth year of the Chunxi era of the Southern Song Dynasty (AD 1179)
(7) The Main Hall of Chen Taiwei Palace in Luoyuan, Fujian Province, built in the third year of the Jiaxi of the Southern Song Dynasty (AD 1239)

The eight wooden buildings of the Liao Dynasty:[37]

(1) The Guanyin Pavilion of Dule Temple in Ji County, Tianjin, built in the second year of the Tonghe era of the Liao Dynasty (AD 984)
(2) The Main Gate of Dule Temple in Ji County, Tianjin, built in the second year of the Tonghe era of the Liao Dynasty (AD 984)
(3) The Main Hall of Fengguo Temple, in Yi County, Liaoning Province, built in the ninth year of the Kaitai era of the Liao Dynasty (AD 1020)
(4) The Bojiajiaocang Hall of Huayan Temple in Datong, Shanxi Province, built in the seventh year of the Chongxi era of the Liao Dynasty (AD 1038)

Main Gate of Dule Temple, Ji County, Tianjin.

(5) The Sakyamuni Pagoda of Fogong Temple in Ying County, Shanxi Province, built in the second year of the Qingning era of the Liao Dynasty (AD 1056)
(6) The Great Hall of Shanhua Temple in Datong, Shanxi Province, inferred to date back to the Liao Dynasty
(7) The Main Hall of Kaishan Temple in Gaobeidian, Hebei Province, inferred to date back to the Liao Dynasty
(8) The Wenshu Hall of Geyuan Temple in Laiyuan, Hebei Province, inferred to date back to the Liao Dynasty

The existing major wooden buildings of the Jin Dynasty:

(1) The Great Hall of Jingtu Temple in Ying County, Shanxi Province, built in the second year of the Tianhui era of the Jin Dynasty (AD 1124)
(2) The Longyan Temple in Lingchuan, Shanxi Province, built in the seventh year of the Tianhui era of the Jin Dynasty (AD 1129)
(3) The Guanyin Hall of Kaihua Temple, in Gaoping, Shanxi Province, built in the first year of the Huangtong era of the Jin Dynasty (AD 1141)
(4) The Rear Hall, East and West Dressing Pavilions of Er'xian Temple in Xixi, Lingchuan, Shanxi Province, built in the second year of the Huangtong era of the Jin Dynasty (AD 1142)
(5) The Meituo Hall and Guanyin Hall of Chongfu Temple in Shuozhou, Shanxi Province, built in the third year of the Huangtong era of the Jin Dynasty (AD 1143)
(6) The Shengmu Hall of Zetian Temple in Wenshui, Shanxi Province, built in the fifth year of the Huangtong era of the Jin Dynasty (AD 1145)
(7) The Main Gate, Tianwang Hall and Sansheng Hall of Shanhua Temple, built in the Tianhui and the Huangtong eras of the Jin Dynasty (AD 1128–1143);
(8) The Puxian Pavilion of Shanhua Temple in Datong, Shanxi Province, built in the second year of the Zhenyuan era of the Jin Dynasty (AD 1154)

TOP Guanyin Pavilion of Dule Temple, Ji County, Tianjin.

BOTTOM Great Buddha Hall of Yanqing Temple, Wutai Mountain, Shanxi Province.

TOP Main Hall of Kaishan Temple, Gaobeidian, Hebei Province.

BOTTOM Main Hall of Chunhua Temple, Pingshun County, Shanxi Province.

Rear Hall of Zhengjue Temple, Zhangzi County, Shanxi Province.

(9) The Middle Hall of Er'xian Temple in Xilimen, Gaoping, Shanxi Province, built in the second year of the Zhenglong era of the Jin Dynasty (AD 1157)

(10) The Main Hall of Dongyue Temple in Yuquan, Lingchuan, Shanxi Province, built in the first year of the Zhenglong era of the Jin Dynasty (AD 1156)

(11) The Main Hall of Zhensheng Temple in Taigu, Shanxi Province, built in the second year of the Zhenglong era of the Jin Dynasty (AD 1157)

(12) The Wenshu Hall of Yanshan Temple in Fanshi, Shanxi Province, built in the third year of the Zhenglong era of the Jin Dynasty (AD 1158)

(13) The Main Hall of Dayun Temple, Qin County, Shanxi Province, built in the Dading era of the Jin Dynasty (AD 1161–1189)

(14) The Main Hall of Fuxiang Temple in Yushe, Shanxi Province, built in the Dading era of the Jin Dynasty (AD 1161–1189)

(15) The Dacheng Hall of Confucian Temple in Pingyao, Shanxi Province, built in the third year of the Dading era of the Jin Dynasty (AD 1163)

Main Gate of Longmen Temple, Pingshun County, Shanxi Province.

(16) The Main Gate of Chengtang Temple in She County, Hebei Province, built in the fourth year of the Dading era of the Jin Dynasty (AD 1164)
(17) The Front Hall of Jinci Temple in Taiyuan, Shanxi Province, built in the eighth year of the Dading era of the Jin Dynasty (AD 1168)
(18) The Main Hall of Chunhua Temple in Pingshun, Shanxi Province, built before the ninth year of the Dading era of the Jin Dynasty (AD 1169)
(19) The Main Hall of the Er'xian Palace in Zhongping, Gaoping, Shanxi Province, built in the twelfth year of the Dading era of the Jin Dynasty (AD 1172)
(20) The Main Hall of Sanzong Temple in Huguan, Shanxi Province, built in the seventeenth year of the Dading era of the Jin Dynasty (AD 1177)
(21) The Dedication Hall of the Dabei Yuan in Quwo, Shanxi Province, built in the twentieth year of the Dading era of the Jin Dynasty (AD 1180)
(22) The South Main Hall of Taiyin Temple, Jiang County, Shanxi Province, built in the twentieth year of the Dading era of the Jin Dynasty (AD 1180)
(23) The Stage of Er'lang Temple in Gaoping, Shanxi Province, built in the twenty-third year of the Dading era of the Jin Dynasty (AD 1183)
(24) Cuifujun Temple Gate in Lingchuan, Shanxi Province, built in the twenty-fourth year of the Dading era of the Jin Dynasty (AD 1184)
(25) The Sanqing Hall of Fengxian Temple in Jiyuan, Henan, built in the twenty-fourth year of the Dading era of the Jin Dynasty (AD 1184)
(26) The Main Hall of Zhaoze Wang Temple in Xiangyuan, Shanxi Province, built in the twenty-seventh year of the Dading era of the Jin Dynasty (AD 1187)
(27) The Dedication Hall and Main Hall of Hutu Temple in Qingxu, Shanxi Province, built in the first year of the Mingchang era of the Jin Dynasty (AD 1190)
(28) The Sakya Hall of Baitai Temple in Xinjiang, Shanxi Province, built during the Mingchang era of the Jin Dynasty (AD 1190–1196);
(29) The Great Hall of Bu'er Temple in Yangqu, Shanxi Province, built in the sixth year of the Mingchang era of the Jin Dynasty (AD 1195)

(30) The Main Hall of Taifu Temple in Fenyang, Shanxi Province, built in the fifth year of the Cheng'an era of the Jin Dynasty (AD 1200)
(31) The Sleeping Quarter of Dawang Temple in Yu County, Shanxi Province, built in the fifth year of the Cheng'an era of the Jin Dynasty (AD 1200)
(32) The Dacheng Hall of Qingyuan Confucian Temple in Qingxu, Shanxi Province, built in the third year of the Taihe era of the Jin Dynasty (AD 1203)
(33) The Main Hall of Prince Lingze Temple in Xiangyuan, Shanxi Province, built in the second year of the Da'an era of the Jin Dynasty (AD 1210)
(34) The Worship Hall of Tangdi Temple in Xiajiao, Yangcheng, Shanxi Province, built in the third year of the Da'an era of the Jin Dynasty (AD 1211)
(35) The Er'lang Temple in Danfeng, Shaanxi, built in the third year of the Da'an era of the Jin Dynasty (AD 1211)
(36) The Main Hall of Qingliang Temple in Dengfeng, Henan Province, built in the fourth year of the Zhenyou era of the Jin Dynasty (AD 1216)
(37) The Sanqing Hall of Huixian Temple in Wuxiang, Shanxi Province, built in the sixth year of the Zhengda era of the Jin Dynasty (AD 1229)

The following are surviving wooden buildings that can be traced back to the Jin Dynasty based on their main architectural features, without an exact year of construction:

(1) The Great Buddha Hall of Yanqing Temple in Wutai, Shanxi Province
(2) The Main Gate of Longmen Temple in Pingshun, Shanxi Province
(3) The Great Hall of Dayun Temple of Jingzhuang Village, Hunyuan County in Datong, Shanxi Province
(4) The Great Hall of Sansheng Temple in Fanshi, Shanxi Province
(5) The Great Hall of Hongfu Temple in Dingxiang, Shanxi Province
(6) The Main Hall of the Buddhist Temple in Chuandi Village, Zezhou, Shanxi Province

(7) The Main Hall of Er'xian Temple in Nanshentou, Lingchuan, Shanxi Province
(8) The Main Hall of Sirunsan Church, Lingchuan, Shanxi Province
(9) The Main Hall of Shizhang Jade Emperor Temple in Lingchuan, Shanxi Province
(10) The Passage Hall of Baiyu Palace in Lingchuan, Shanxi Province, built during the Da'an and the Chongqing period of the Jin Dynasty
(11) The Main Hall of Beima Jade Emperor Temple in Lingchuan, Shanxi Province
(12) The Main Hall of Hongji Yard in Wuxiang, Shanxi Province
(13) The Great Hall of Xiangyan Temple in Liulin, Shanxi Province
(14) The Main Hall of Xingdongyuan Dongyue Temple in Shilou, Shanxi Province
(15) The Dedication Hall of Kaifu Temple in Yangcheng, Shanxi Province
(16) The Main Hall of Longwang Temple in Dongyi, Lucheng, Shanxi Province
(17) The Middle and Rear Halls of Tianwang Temple in Changzi, Shanxi Province
(18) The Rear Hall of Zhengjue Temple in Changzi, Shanxi Province
(19) The Main Hall of Cixiang Temple in Pingyao, Shanxi Province
(20) The Zhongfo Hall of Fengxue Temple in Ruzhou, Henan Province
(21) The Pilu Hall and Great Hall of Lingshan Temple in Yiyang, Henan Province

During the Song, Liao, and Jin dynasties, important changes occurred in both cities and buildings, which was a turning point where ancient Chinese architecture evolved from its mature state to a more complicated, delicate, and diverse style. The development of Neo-Confucianism in the Song Dynasty guided social ideologies and had an important influence on the artistic style and aesthetics of architecture, such as the order of ritual and ethical cultivation that appeared in architecture. The overall architectural style pursues delicacy and softness. The early form of cities began to transform into the modern style, where cities became street-oriented and commercialized. In the same period, the social, economic, and cultural development led to some new types of commercial, entertainment, and educational facilities such as theaters and colleges. With the development of brick-burning technology, masonry was consumed in large quantities, and the carving style also became very splendid and exquisite. In the Song, Liao and Jin dynasties, architectural color painting was widely used, resulting in increasingly complicated architectural decoration. The system of brackets and beams became more mature and delicate, and the style became more graceful.

During this period, the second year of the Chongning era of the Northern Song Dynasty (AD 1103) was an extremely important year, during which the *Yingzao Fashi*, compiled by the young Supervisor of Works Li Mingzhong, was published. It is the first architectural technical standard in history and a summary of long-standing technical practices in the Chinese construction industry.[38] The completion of the book allows us to have a deeper and more detailed understanding of the buildings from the Song, Liao, and Jin dynasties and the later architectural developments based on the remains of the period. It also provides our inference of early buildings with the coordinates of critical references to guide us like a lighthouse on a night voyage. In addition, the *Mu Jing*, written by Yu Hao, the famous craftsman of the Northern Song Dynasty, has now been lost. A limited description of the book can only be found in *Mengxi Bitan* (*Dream Pool Essays*).

The buildings of the Liao Dynasty still largely retained the architectural style of the Tang Dynasty, despite the strong influence of the Song Dynasty architecture of the same period. Many Liao buildings featured a simple rectangular floor plan topped by a four-cornered pavilion roof. They adopted the Tang Dynasty style of using wooden plank doors and straight lattice windows. The overall architectural style was robust, solid, and unadorned.

The Jin Dynasty, which coexisted with the Southern Song Dynasty, comprehensively developed the characteristics of Song and Liao architecture on the basis of the Liao Dynasty model. The architectural style focuses on creativity and changes in details. The quality became more delicate and gradually caused a cumbersome production process.

Decline: Yuan, Ming, and Qing Dynasties

The founders of the Yuan Dynasty were originally Mongolian nomads in outer Mongolia. In AD 1206, Genghis Khan (the founder of Yuan) ascended the throne as the Great Khan. Emperor Taizong of Yuan destroyed Jin in AD 1234, and Kublai Khan (Emperor Shizu of Yuan) founded the Yuan Dynasty in AD 1271. In AD 1279, he ended the Southern Song Dynasty and unified China.

After the Red Turban Rebellion at the end of the Yuan Dynasty, Zhu Yuanzhang (Emperor Taizu of Ming) established the Ming Dynasty in AD 1368. After 267 years and 16 emperors, it was overthrown by a peasant uprising led by Li Zicheng in AD 1644.

In the same year, the Manchu Qing entered the customs to establish the Qing Dynasty. After 268 years, 12 emperors were passed on. In AD 1912, Emperor Puyi of the Qing Dynasty abdicated, marking the end of the Qing Dynasty.

The Yuan Dynasty occupies a special place in Chinese history. On the one hand, it ended the division of the Song, Liao, Jin, and Xia regimes and achieved great reunification; on the other hand, the reunification was achieved through violence, which severely damaged the society and the economy and seriously affected agriculture and industry. During the Yuan reign, dual oppression existed in terms of both class and ethnicity, which greatly hindered the development of architecture greatly. It was not until the Kublai Khan adopted a policy of promoting agriculture and sericulture that social productivity was gradually restored. Previously nomadic, the Yuan people used tents as their basic dwelling, and thus did not adhere to the architectural forms in the areas in which they ruled. This allowed the free development of all kinds of buildings and promoted the exchange of architectural forms in different regions to a certain extent, which led to the formation of the fundamental pattern of Yuan Dynasty architecture. Some technological innovations appeared, such as Yingding and other design changes.

The construction of Dadu is a typical example of the urban planning and architectural achievements of the Yuan Dynasty.[39] The Dadu of Yuan was designed according to the layout of the traditional capital of the Han and took eight years to build. The main palace is located at the southern end of the city's central axis. On the east and west sides of the Imperial City, the Imperial Ancestral Temple and the Altar of Land and Grain were respectively constructed following the traditional layout of "Ancestral Temple on the left and Earth God Shrine on the right." Various religions coexisted during the Yuan Dynasty, and many temples were built in Dadu, including Huguo Temple, Miaoying Temple, Dongyue Temple, and many others. During this period, Lamaism, which was popular in Tibet, became known in mainland China with the construction of some Tibetan Buddhist buildings. Tibetan pagodas became one of the important types of Buddhist pagodas. Wooden buildings still followed the construction practices of the Song and Jin dynasties, but they were inferior in scale and quality to those of the previous dynasties. The size of beams and brackets was much smaller than in the Song and Jin dynasties. However, the design also became simpler and grander than during the Ming and Qing dynasties. Some buildings were built without interior beams and brackets, while others adopted a simpler method called "Da'e style," where the beams and brackets are replaced by a single thick head instead of resting on the main vertical columns and column heads. Due to the prevalence of architectural techniques such as the reduction and relocation of columns in previous dynasties, curved wood can also be used as beams without cutting to present its original state. Additionally, roofs were built to be exposed and covered, often without additional ceiling ornamentation. The building materials were also used in a relatively hasty and rough manner as a result of the economy measures taken due to the slow economic development. The economic method also had another effect. The design of the building pays more attention to the integrity and stability of the structures, simplifying some components to make the structure clearer while maintaining its solidity. There are many remnants of wooden buildings from the Yuan Dynasty, the most famous of which are the Yongle Palace in Ruicheng, Shanxi Province, the Guangsheng Temple in Zhaocheng, and the Qinglong Temple in Jishan (the murals in the Yongle Palace and the Guangsheng Temple are considered to be models for the murals of the Yuan Dynasty); the Ciyun Pavilion in Dingxing, Hebei, the Yanghe Building in Zhengding, the Dening Hall in Quyang Beiyue Temple, and the Shenggu Temple in An'ping. The more typical wooden buildings of the Yuan Dynasty in Jiangnan (the region south of the Yangtze River) include the Zhenru Temple in Shanghai; the Tianning Temple in Jinhua, Zhejiang, the Yanfu Temple in Wuyi, the Great Hall of the Shisi Temple in Jingning, Lishui; the Xuanyuan Palace in Dongshan, Suzhou, Jiangsu. The earliest surviving dwellings from this period are the Jishi Dwellings in Shanxi Province.

TOP Main Hall of Cuifujun Temple, Zhangzi County, Shanxi Province.

BOTTOM Xuanyuan Palace, Suzhou, Jiangsu Province.

Main Hall of Zhenru Temple, Shanghai.

TOP Sanzong Temple, Dazhonghan Village, Zhangzi County, Shanxi Province.

BOTTOM Residential house, Zhongzhuang Village, Gaoping, Shanxi Province, Yuan Dynasty.

TOP Residential house, Nanyang Village, Gaoping, Shanxi Province, Yuan Dynasty.

BOTTOM Lingkuangwang Temple, Ju Village, Zhangzi County, Shanxi Province.

The development of brick technology during the Ming Dynasty was particularly remarkable. Before the Yuan Dynasty, ancient wooden buildings were mainly made of mud walls. Bricks were mainly used for foundation walls, platforms and pavements. After the Ming Dynasty, brick walls became common. The adoption of brick walls also led to the rapid development of brick architecture, resulting in the wide use of brick carving decoration and embellishment. In the Ming Dynasty, components such as glazed bricks and tiles reached an unprecedented level of quality, with a wide variety of colors, exquisite patterns, and excellent texture. In addition to palaces and temples, glazed bricks and tiles were also used in pagodas, doors, and walls. After the simplification during the Yuan Dynasty, wooden buildings evolved towards stronger structural integrity, decorative beams and brackets, and streamlined construction processes. In the construction of pavilion buildings, a column-and-beam structure connected from the ground to the roof is used instead of a stacked bracket structure; the penetrating tie is widely used, in which the bracket arms are extended to support the eaves directly while the bracket system at the column head was no longer valued for its structural function, "ang," originally used as an oblique beam, has also become a decorative component.[40] The connections between the roof trusses were simplified to longitudinal components such as purlins, padding blocks, and rafters, and the three previous longitudinal connection methods of purlins, climbers, and strings adopted before the Song Dynasty were no longer used. Song Dynasty practices such as raised column bases, corner brackets at the eaves of columns, brackets at the top of columns, and moon-shaped beams were also gradually eliminated, except for some dwellings in Jiangnan (south of the Yangtze River). The official wooden buildings of the Ming Dynasty were generally austere and stable in style, possessing both the grand style of the Tang Dynasty and the gorgeous and exquisite style of the Song Dynasty, forming a unique style that was characteristic of its era.

Chinese architecture matured during the Qing Dynasty. The system of government construction management tended to be standardized, but the design became more rigid. Residential buildings were rich and varied, and many buildings of different styles were created in different regions and ethnic groups. Tibetan Buddhist architecture flourished during this period. Various architectural forms were created on the basis of traditional Buddhist architecture. Garden architecture

TOP Paiyun Hall, Summer Palace, Beijing.

BOTTOM Yuhua Hall, Summer Palace, Beijing.

flourished during the Qing Dynasty, especially in the south of the Yangtze River. The garden architecture was widely constructed at a high technical level. The royal gardens and private gardens of the Ming and Qing dynasties have both reached an unprecedented artistic level. The architecture of the Qing Dynasty stood out the most in terms of the aesthetics of architectural decoration. The northward migration of the Jiangnan architectural craftsmanship has greatly improved the decoration of northern buildings, including wooden lattice, brick and wood carving decoration, jade and stone inlays, landscape gardens, and moon-shaped floor inlays. The technical level of stone carving, brick carving, and wood carving in the Qing Dynasty was quite high, with outstanding techniques such as bas-relief, openwork carving, stacked carving, enclosed carving, hollow carving and inlaid carving; decorative elements incorporated the use of inlaid jewelry and jade, bamboo, wood, seashells and cloisonne. Architectural color painting also became a vital part of architectural decoration during the Qing Dynasty. The building structure was almost liberated from the constraints of the bracket system, becoming primarily a decorative structure. The connections between beams, rafters, columns, and purlins were made by direct mortise-and-tenon joinery, which efficiently transferred stress, reduced material consumption, and enhanced structural stability and balance. Beams and rafters were connected in a flexible manner, which led to the creation of distinctive architectural characteristics of the Qing Dynasty, such as various roof structures and overlapping roofs. The overall architectural style was characterized by a sense of dignity, grandeur, and precision.

The labor system of palace construction in the Qing Dynasty underwent an important change. It shifted from a system reliant on military and conscripted labor to one based on wage labor and private enterprise. This transition gave rise to the famous families of craftsmen of Qing Dynasty palace architecture, the "Lei Family" Lei Fada and Lei Jinyu. Every single architectural design plan of the Lei family was followed by an initial model at a scale of 1/100 or 1/200 for the approval of the Inner Chamber. These models are called hot-water bending models, which are used both as building construction plans and as specifications for the corresponding types of buildings and are of great value for the study of Qing Dynasty architecture.

In addition to *Yingzao Fashi*, another book on architectural techniques was published, the *Gongcheng Zuofa Zeli*, which was promulgated by the government in AD 1733. It marked a period of highly standardized, typified, and institutionalized architecture in the Qing Dynasty, as well as a high level of maturity in the structural system of Chinese wooden buildings.

NOTES

1. Li Yunhe, *Principles of Ancient Chinese Architectural Design* (Tianjin: Tianjin University Press, 2005), 21.
2. Liang Sicheng, *The Complete Works of Liang Sicheng*, vol. 4 (Beijing: China Architecture & Building Press, 2001), 16.
3. Liu Dunzhen, *The Complete Works of Liu Dunzhen*, vol. 6 (Beijing: China Architecture & Building Press, 2007), 8.
4. Fu Xinian, *An Overview of Ancient Chinese Architecture* (Beijing: Beijing Publishing House, 2016), 3.
5. Sun Dazhang, *A Brief History of Ancient Chinese Architecture* (Beijing: Tsinghua University Press, 2016), 3.
6. Pan Guxi, *The History of Chinese Architecture* (Beijing: China Architecture & Building Press, 2015), 17.
7. Li Yunhe, *Principles of Ancient Chinese Architectural Design* (Tianjin: Tianjin University Press, 2005), 22.
8. Chen Wei, *Ten Books by Contemporary Chinese Architectural Historians (Selected Essays on Architectural History by Chen Wei)* (Shenyang: Liaoning Fine Arts Publishing House, 2015), 328.
9. Li Yunhe, *Principles of Ancient Chinese Architectural Design* (Tianjin: Tianjin University Press, 2005), 23.
10. Liu Xujie, *The History of Ancient Chinese Architecture*, vol. 1 (Beijing: China Architecture & Building Press, 2009), 30.
11. Pan Guxi, *The History of Chinese Architecture* (Beijing: China Architecture & Building Press, 2015), 19.
12. Ibid., 26.
13. Fu Xinian, *An Overview of Ancient Chinese Architecture* (Beijing: Beijing Publishing House, 2016), 4.
14. Liu Xujie, *The History of Ancient Chinese Architecture*, vol. 1 (Beijing: China Architecture & Building Press, 2009), 617.

15. Remains of Han Dynasty pottery houses and Han *que*, as well as many Han tombs, are considered important examples for studying the architectural forms of Han wooden structures, although they use different materials than wooden structures.
16. Liu Dunzhen, *The Complete Works of Liu Dunzhen*, vol. 6 (Beijing: China Architecture & Building Press, 2007), 24.
17. Fu Xinian, *The History of Ancient Chinese Architecture*, vol. 2 (Beijing: China Architecture & Building Press, 2009), 2.
18. Ibid., 42.
19. Liu Dunzhen, *The Complete Works of Liu Dunzhen*, vol. 6 (Beijing: China Architecture & Building Press, 2007), 38.
20. In AD 386, the Xianbei tribe Tuoba Gui founded a country in the present-day northern Shanxi Province and Inner Mongolia. In AD 398, the country was renamed Wei, with its capital at Pingcheng (modern-day Datong). In AD 439, Northern Wei unified the north, and in AD 493, the capital was relocated to Luoyang.
21. The Qianfo Stone Pagoda of the Northern Wei Dynasty, also known as the "Cao Tiandu Stone Pagoda," was constructed in the first year of the Tian'an era of the Northern Wei Dynasty (AD 466) by the imperial attendant Cao Tiandu for the purpose of praying for blessings and mourning deceased relatives. The stone pagoda, contemporaneous with the Yungang Grottoes, is the earliest known extant statue pagoda in China. The Taipei Historical Museum houses the pagoda's body and base, while the Mayi Museum displays a replica of the pagoda spire.
22. Currently housed in the Museum of Fine Arts, Boston, USA.
23. This is the earliest extant stone pagoda in China and the earliest pagoda in the pavilion style.
24. The largest ancient tomb pagoda in China and the earliest brick imitation wooden tower.
25. The Great Wild Goose Pagoda was initially built in the third year of the Yonghui era (AD 652) of the Tang Dynasty. It collapsed and was rebuilt during the reign of Empress Wu Zetian in Chang'an (AD 701–704). The current appearance of the pagoda's exterior reflects the addition of outer walls during the Ming Wanli era. The lintel of the western gate on the ground floor is engraved with a Buddha hall image, reflecting the early Tang architectural style, making it a valuable resource in architectural history.
26. The precise dating is unclear, but in the Ming Dynasty, the scholar Xu Bangcai's *Record of the Reconstruction of Jiuta Temple* suggests it may be from the Tianbao era, while Liang Sicheng believed it to be from the Tang Dynasty. In his book Ancient Pagodas in China, Luo Zhewen further identified it as belonging to the Tianbao period of the Tang Dynasty.
27. The pagoda dates back to the first year of the Bao'en era of Emperor Xuanzong of the Tang Dynasty (AD 762), making it the only brick relief pagoda.

28. The Lamp-Burning Pagoda, also known as the Changming Lamp, is one of six Buddhist offerings, also known as lamp stands or lamp platforms.
29. The earliest known example of a round tomb pagoda and a unique example of a Tang round pagoda.
30. The earliest extant iron pagoda, with a precise dating.
31. In 1933, Liang Sicheng and Lin Huiyin conducted a survey of the Bell Tower of Kaiyuan Temple. In his article "Brief Notes on the Investigation of Zhengding," Liang stated, "If it is a Tang structure, I cannot deny it." Subsequent experts discovered a complete Tang architectural structure on the first floor, while the second floor was altered during the Qing Dynasty. In the late 1980s, the Bell Tower underwent significant renovations, with the first floor retaining its original form and the second floor restored to its Tang Dynasty style.
32. When Tiantai Temple was listed as a national cultural heritage site in 1988, it was dated to the Tang Dynasty. During a major renovation in 2014, an inscription reading "Great Tang Tiancheng" was discovered on the main hall's roof trusses, leading Tang Dahua to infer it was from the Tiancheng era of the Later Tang Dynasty.
33. In 1933, Liang Sicheng believed that the ancient buildings he examined in Zhengding were relics from the late Tang and Five Dynasties periods. Luo Zhewen believed that this hall should be the earliest surviving Confucian temple main hall in China.
34. Liu Dunzhen, *The Complete Works of Liu Dunzhen*, vol. 6 (Beijing: China Architecture & Building Press, 2007), 70.
35. The Wu Yue Kingdom of the Five Dynasties period corresponds to the second year of the Qiande era of the Northern Song Dynasty (AD 964).
36. Early wooden ancient buildings in Jiangnan during the Yuan Dynasty include the main hall of Yanfu Temple in Wuyi, Zhejiang (AD 1317, the fourth year of Yanyou era), the main hall of Tianning Temple in Jinhua, Zhejiang (AD 1318, the fifth year of Yanyou era), the main hall of Zhenru Temple in Putuo District, Shanghai (AD 1320, the seventh year of Yanyou era), and the two main gates of Huqiu in Suzhou, Jiangsu (AD 1338, the fourth year of Zhiyuan era).
37. The initial list of "Eight Great Liao Structures" included the Three Great Officials Hall of Guangji Temple in Baodi District, Tianjin, and the Puxian Pavilion of Shanhua Temple in Datong, Shanxi Province. The Three Great Officials Hall of Guangji Temple was demolished in AD 1947 (and began to be reconstructed in an ancient style in 2005). Following the founding of the People's Republic of China, the Puxian Pavilion of Shanhua Temple discovered inscriptions from the Jin Dynasty during a major renovation, leading these two buildings to be removed from the Liao structures. Later, two additional Liao-era buildings were discovered: the Kaishan Temple in Gaobeidian, Hebei, and the Wenshu Hall of Geyuan Temple in Laiyuan, Hebei, which further expanded the list.

38. Guo Daiheng, *The History of Ancient Chinese Architecture*, vol. 3 (Beijing: China Architecture & Building Press, 2009), 6.
39. Pan Guxi, *The History of Ancient Chinese Architecture*, vol. 4 (Beijing: China Architecture & Building Press, 2009), 3.
40. Pan Guxi, *The History of Chinese Architecture* (Beijing: China Architecture & Building Press, 2015), 49.

CHAPTER 3

PLATFORM

Every House Has Three Parts

Gaining an understanding of ancient Chinese architecture is akin to getting to know a person. It is essential to gain an overall impression and visual experience by looking up and down. Similarly, in architecture, this vision is derived from the form and style of the architecture itself. A profound understanding can be achieved at that moment.

In Volume 18 of *Mengxi Bitan*, Shen Kuo of the Northern Song Dynasty records that Yu Hao, a craftsman of the Song Dynasty, wrote *Mu Jing*, which states, "Every house is structured into three parts." The upper part is composed of beams and a frame above, with the middle section between the beams and the ground, and the steps and under constitute the lower part.[1] Yu Hao's focus was on the three divisions of a house, from the roof to the platform, with the body in between.

Mr. Liang Sicheng provided a more detailed overview of the topic, "Chinese architecture is divided into three distinct categories based on its three-dimensional layout: The three main components of Chinese architecture are: (1) platform, (2) wall column frame, and (3) roof. Regardless of the location, dynasty, or function of a building, the three components are present. If the structure includes splendid red pillars and exquisitely painted foreheads crisscrossing, it is likely that glazed tiles will be present on the crown-like top. The base may consist of a single or multi-layer masonry pedestal to support the structure. For thousands of years, the different materials, functions, and structures of the tripartite have been seamlessly integrated into the layout of one building. Any emphasis on a particular component would never diminish the structural and functional significance of any other parts."[2]

The three-part perspective aligns with a fundamental observation principle. When observing mountains or trees, individuals typically follow a hierarchical order, starting with the summit, then the middle, and finally the base. This approach is not exclusive to the field of architecture or Chinese buildings. For all architectural styles, including Western ones, the three-part division is a common approach to observing elevations. However, Western architectural styles tend to prioritize the integrity of the elevation more than Chinese architectural styles. Nevertheless, the three-part theory, which is neither a feature of the elevation

Every house is structured into three parts.

composition nor a theoretical doctrine in Chinese architecture, has long been a standard to distinguish ancient Chinese buildings from other architectural systems.

Mr. Li Yunhe proceeded to elaborate further on this three-part theory. He made the following assertion, "I believe this three-part theory exerts a wide influence because in history, most books on traditional Chinese architecture, domestic and foreign, have used this theory as a point of departure. Rather than focusing only on the three parts, Chinese classical architecture is more concerned with the equal treatment of every part in roughly equal proportions and in respective functions and shapes. This conscious approach puts them in a contrasting state with no intention to integrate into a whole the tripartite that works and builds in sharply different ways."[3]

The perspectives of observation play a critical role in this three-part theory. The height of Chinese architecture determines a long-distance perspective, so the roof is shaped and decorated with special importance. The mid-distance viewing requires a harmonious and appropriate proportion of the tripartite in one building, which differs greatly from Western architecture that focuses on the facade wall, the bottom, and the top, all of which vary little except for their functional changes. However, since most of the facades of Chinese houses are courtyards and interior walls, decoration from a close perspective and smooth spatial transitions are emphasized,[4] so with little changes in the facade wall, the focus of construction is cast on variations in roof and platform.

Mid-distance perspective in architecture.

Creation of Platform Bases

A platform base is a highly developed ingredient in ancient Chinese architecture. As Mr. Liang Sicheng stated, "A platform base is the foundation of all buildings and a defining feature of Chinese architecture. While similar platform shapes are also found in European architecture, their function is not as essential as in Chinese buildings."[5] As the levels of construction increase, platforms vary in size and design. As the level of construction increases, the foundation becomes taller and larger, with some bases even comprising multiple layers. For tall and multi-layered platforms, railings are erected to satisfy functional and secure demands, creating a distinctive platform landscape for Chinese architecture.

When was the platform first introduced? The *Mozi · Ciguo* states, "Prior to the development of the ability to construct palaces, ancient people lived in mausoleums and caves. The water below was a source of discomfort, prompting the sage king to initiate the construction of palaces. The construction of a palace involved ensuring the height of the building to prevent moisture, the sides to resist wind and cold, and the top to receive snow, frost, rain, and dew." In essence, when the ancient people lacked the knowledge to construct dwellings, they resided in caves on elevated terrain. Given the adverse health effects of moisture in the ground, ancient saints initiated the construction of houses. The construction of houses involved raising the ground to prevent moisture, surrounding all sides to resist wind and cold, and building roofs to protect houses from falling snow, frost, rain, and dew. The historical records demonstrate that from the inception of house construction, a platform or a structure analogous to a raised ground was utilized.

Examples of platform foundations emerged as early as the Neolithic period. Among the Qujialing cultural and Chengtoushan cultural relics,[6] some of the well-preserved, high-standard, and large-scale platforms were concentrated on the three houses near the center of the city site. All of the foundations were constructed using loess platform foundations with slopes on all four sides. Prior to construction, foundation trenches were dug on the platform surface, and the living area was trimmed. The plan shape was square or rectangular.[7] Archaeological excavations at different sites in the Xia, Shang, and Zhou dynasties and later generations also evidenced the intentional construction of rammed earth platform bases. Some statistics of these findings are listed below: in the Erlitou site in Yanshi, Henan Province,[8] the height of the platform bases at the No. 1 palace site was 0.8 meters, while 0.25–0.4 meters at the No. 4 site and 0.1–0.3 meters at No. 5, respectively. In the Zhengzhou Shangcheng site, the platform bases were generally 1 to 2 meters high.[9] At the Yinxu palace buildings, the platform bases reached as high as 1 meter.

Different from Mozi's statements that ancestors built the platform bases to avoid "moistening and harming the people" in house construction, Mr. Li Yunhe proposed that a platform was constructed as a safety device to prevent floods and waterlogging since the Yellow River Basin is the birthplace of the Han nationality and the Yellow River frequently flooded ever since ancient times. "At least this supports the fact that the Chinese faced significant flooding challenges in the 20th and 30th centuries BC. After experiencing these floods, the Chinese developed solutions to protect their homes, including raising them above the ground or constructing platforms above the surrounding area."[10] In addition, these platforms also played a defensive role in wars. Mr. Li Yunhe concluded that a platform serves two main purposes: flood and waterlogging prevention, as well as war defense. All in all, Mr. Li Yunhe summed up two functions of a platform:

flood and waterlogging prevention and war defense. Additionally, it represents the owner's status and power.

However, Mr. Li Yunhe's conclusion is open to question. The first function of preventing flood and waterlogging is dubious, as evidenced by the Dayu example. Given that the Yellow River Basin is the birthplace of the Han people, it is evident that the prevalent way to control floods involved the construction of dams and the digging of river channels rather than the use of a lifted platform. A platform base was at best moisture-proof and waterlogging-proof. In fact, it is likely Mr. Li Yunhe confused the meaning of "flood" with "waterlogging" since "waterlogging" refers to the soaking caused by continuous heavy rain failing to be drained out in time, but an elevated foundation could only resist ground tides and rainwater immersion. Therefore, a platform foundation designed for flood control is unconvincing. In addition, the idea of platforms occupying a commanding height for war defense is impractical without strong support.

It is evident that platforms designed to fulfill functional needs also provided a solid foundation for walls and columns to stabilize a house. Furthermore, the prevalence of high-rise architecture from the Warring States Period to the Western Han Dynasty saw platforms endowed with ritual norms and spiritual orientation, symbolizing the identity and status of a house owner as well as a contributor to a harmonious, holistic vision of a house.

As evidenced by historical records, the term *tang* was used at least as early as the Qin and Han dynasties to refer to a platform base.

The Book of Rites · Ritual Vessels stated, "Some individuals prioritize height. The emperor's *tang* is nine feet high, the princes' seven feet high, the officials' five feet high, and the scholars' three feet high." *The Rites of Zhou · Winter Officials* recorded, "*Mingtang* (a hall where imperial edicts were proclaimed) in Zhou Dynasty was measured by *yan* (a bamboo mat) of nine feet long, nine *yan* of east and west, seven *yan* of north and south, one *yan* of *tang*." "For *shishi* (with the same function as *mingtang*) of the Xia Dynasty, the depth of the front and back of the *tang* was seven *bu* (steps), and the width was four times the depth [28 *bus* (steps)]... For *chongwu* (with the same function as *mingtang* and *shishi*) in the Yin Dynasty, the depth of *tang* was seven *xun*, the height three feet, hip-roofed and double-eaved." In literature records, the foundations of the emperor's buildings, including *mingtang*, *shishi*, and *chongwu* are all titled *tang*.[11]

TOP Single platform engraved in the portrait bricks of a tomb in the Eastern Han Dynasty on Yangzi Hill, Chengdu.

MIDDLE Stone-engraved wooden platform in Ning Mao's Stone Chamber of the Northern Wei Dynasty.

BOTTOM Double platforms depicted in stone engravings on the tympanum of Dayan Pagoda, Xi'an, Shaanxi Province.

In the Song Dynasty, platform bases were called *jieji*. *Yingzao Fashi* states, "There are four names for platforms: *jie*, *bi*, *gai*, and *shang*."

In the Qing Dynasty, platform bases were named *taiji*. According to *Daqing Huidian Shili* (*Examples of the Qing Dynasty Ceremony*), "The platform for the houses of the third rank below dukes and princes is two feet high, and the fourth rank and the common people is one foot high."

When did the word *taiji* begin to be used for a platform base? From a review of historical literature, it is not clear which dynasty first used the term "taiji." In addition, from the Warring States to the Qin and Han dynasties, high platforms were constructed extensively for the observation of celestial phenomena and the awaiting of immortals. "Platforms could be designated as *tai* if utilized for elevated observation." Therefore, the concept of platforms reached a consensus at that period as *a high ground for observations*. However, there is no evidence that the concept of the platform was applied to *a house foundation* at the end of this period. What is reflected in literature is that before the Qin and Han dynasties, the term *tang* was used to refer to a platform base in the emperor's *mingtang*. However, further study is needed to understand the relationship and distinction between *tang* as a platform and *tang* in *mingtang* (a hall where imperial edicts were proclaimed). Fortunately, by the Qing Dynasty, the documentary records had become relatively unified in naming platform bases as *taiji*.

A platform is typically comprised of a base to support the body (also known as the foundation), steps for people to ascend and descend (also called *taijie*, *tabu*, *caiduo*), and railings surrounding higher platforms for safety purposes.

Platform Bases

In order to achieve a harmonious aesthetic balance for a building, it is important to consider the relationship between the height of the house and its base. If the house is majestic and tall, especially with a large roof, it is essential to ensure that the house base is also tall to avoid top-heaviness. However, this tallness will naturally render the house cramped in space. Consequently, a platform is incorporated at the front of the house base. A house base is comprised of two components: a platform and a terrace. The majority of platforms feature steps on three sides, while smaller platforms typically have steps on the front. A platform expands the outdoor secondary space, enhances the architectural design of the house, and elevates the building's grade and status.

From the initial discovery of the Chengtoushan site, which is believed to be the earliest Neolithic site, platform bases were filled with rammed earth. As Lao Tzu said, "A nine-story platform begins with plain rammed earth." The platform bases of architectural sites in the Shang Dynasty were constructed using a mixture of plain soil, lime earth, and sandstone, a technique that was later

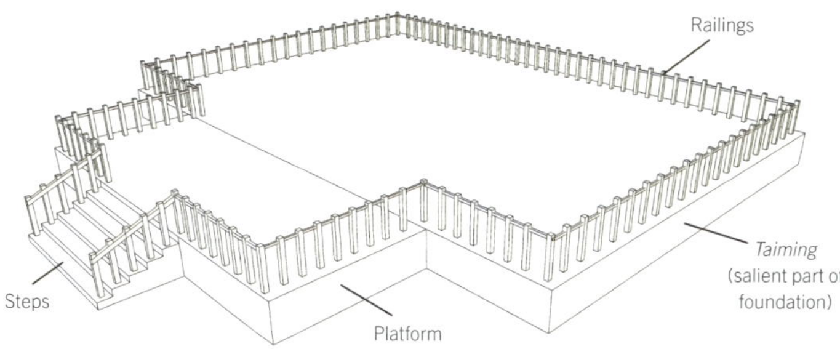

TOP Platform bases.

BOTTOM Ming *xumizuo* platform, Lingkuangwang Temple, Ju Village, Zhangzi County, Shanxi Province.

adopted by subsequent generations. In the Eastern Han Dynasty, a platform foundation began to apply rammed earth inside and masonry outside, adopting a similar shape to that seen today. It is possible that this type of foundation may have emerged earlier. In contrast to the long-standing tradition of platform bases constructed from plain rammed earth, which continued to be used in low-grade or simple houses well into the later periods, the date of the emergence of platform foundations made of mixed rammed earth and masonry wrappings remains uncertain.

In the construction of the Longshan culture period, fired white lime was utilized as a raw material. At the same time, adobe bricks were discovered. Archaeological excavations in the Western Zhou Dynasty revealed the use of tiles and square bricks for paving the ground. By the Spring and Autumn Period, brick technology had reached a mature stage. The earliest evidence of wooden buildings in China dates back to the Tang Dynasty. Prior to this, the only records of house construction were images or architectural ruins. This makes it difficult to determine the exact periods when certain construction technologies were used. In the Eastern Han Dynasty, images of engraved relics clearly showed platform foundations of rammed earth and outer masonry, as well as bricks and white ash that had been used earlier. The findings confirm that this type of platform base was used by the Eastern Han Dynasty.

The architectural remains of the Tang Dynasty exhibited the shape and structural materials of platform bases, and by Song, "standardized" regulations came into being for platform constructions, as illustrated specifically and concretely in *Yingzao Fashi* regarding the technical, material, and grade indexes.

Platform bases can be divided into two categories: ordinary bases and *xumizuo* bases (Sumera seats).

Ordinary bases

Ordinary bases are more commonly used in gardens and buildings of general grade or for residents.[12] Due to the rarity of plain rammed earth foundations in later generations and the prevalence of rammed earth foundations covered with masonry since the Han Dynasty, ordinary bases are typically understood to refer to the latter type. The masonry wrapping for square-shaped ordinary bases is comprised of three main elements: rectangular stone slabs,[13] intermediate piers,

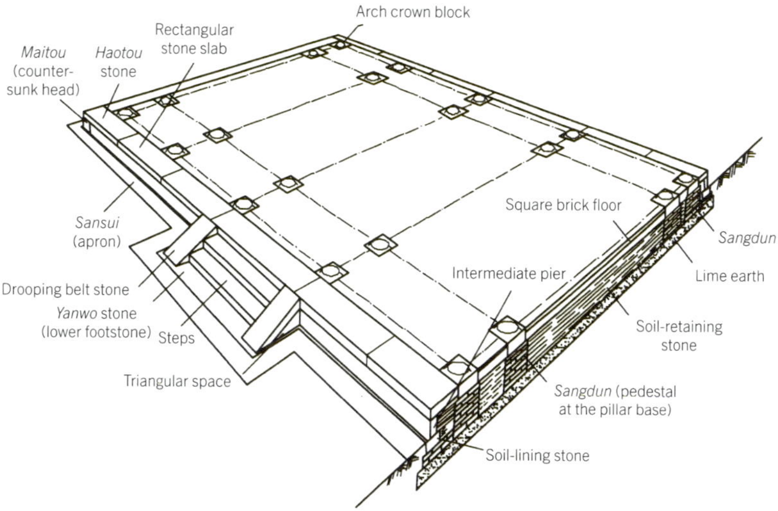

TOP The composition of an ordinary base.

BOTTOM Platform of the Main Hall of Kaihua Temple, Gaoping City, Shanxi Province.

and soil-lining stones. These elements have been used in a variety of forms and applications across different eras and regions.

Rectangular stone slabs are strips of stones tiled along the top surface of a base as a structural transition between the elevation and the plane. For different locations, they are named differently; for example, *haotou* stones at the four corners, while *yushan* strip stones are employed on the side of the gable. For the sake of simplicity, all are referred to as rectangular stone slabs.

Intermediate piers are also named *douban* stones, lying between rectangular stone slabs and the soil lining stones. The lower end is installed in the trough of the soil lining stone, while the upper end is made into a tenon and inserted into the tenon socket under the rectangular stone slabs. The majority of the intermediate piers utilize either stone or brick, the former named *manzhuang* (full stone seats), the latter called "brick *taiming*."

TOP LEFT Platform of the Side Hall of Yuhuang Temple, Bu Village, Zhangzi County, Shanxi Province.

TOP RIGHT Platform of the Yuanjue Hall of Faxing Temple, Zhangzi County, Shanxi Province.

BOTTOM LEFT Platform of Nanchan Temple Wutai Mountain, Shanxi Province.

BOTTOM RIGHT Platform of the Bell Tower of Kaiyuan Temple, Zhengding County, Hebei Province.

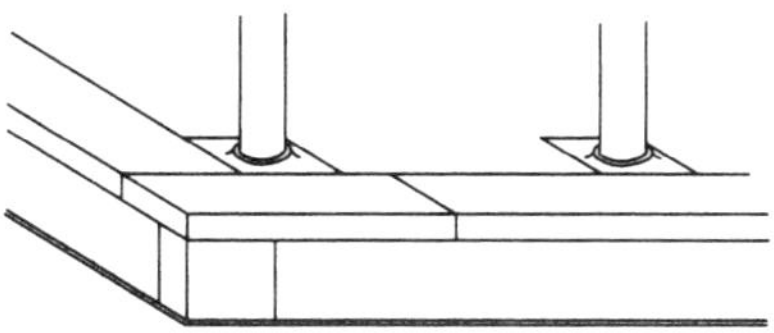

All stone—*douban taiming*

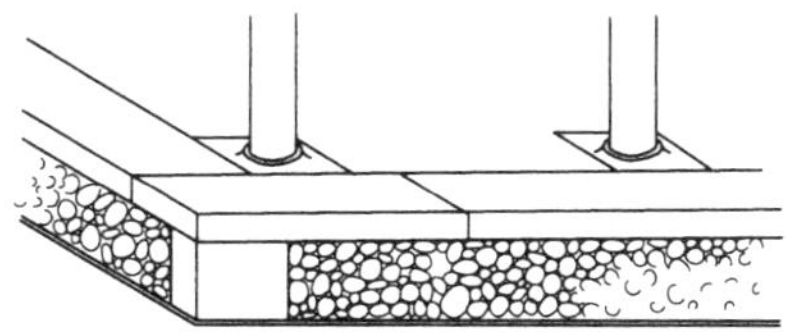

All stone—boulder *taiming*

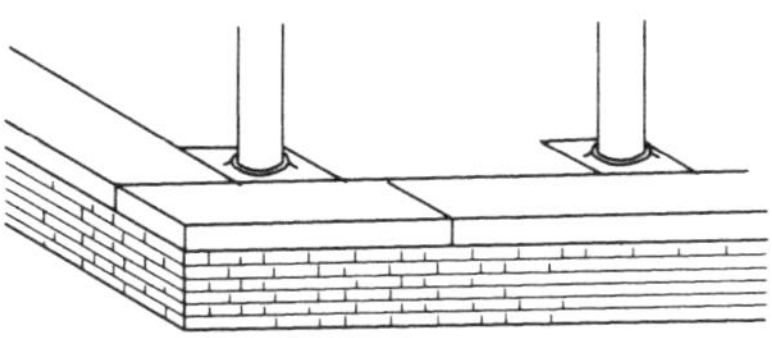

Brick-and-stone—brick *taiming*

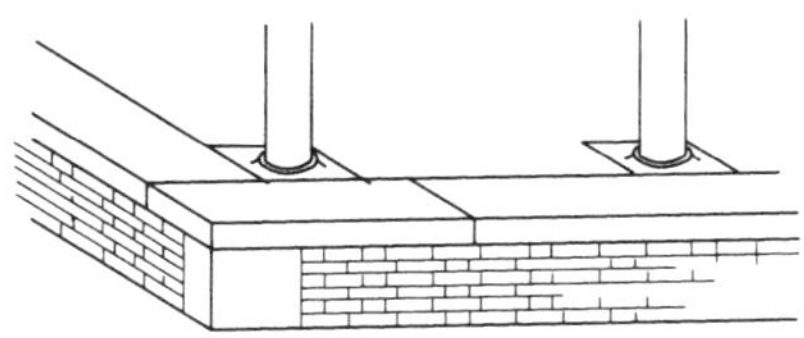

Brick-and-stone—brick *taiming* & stone corner post

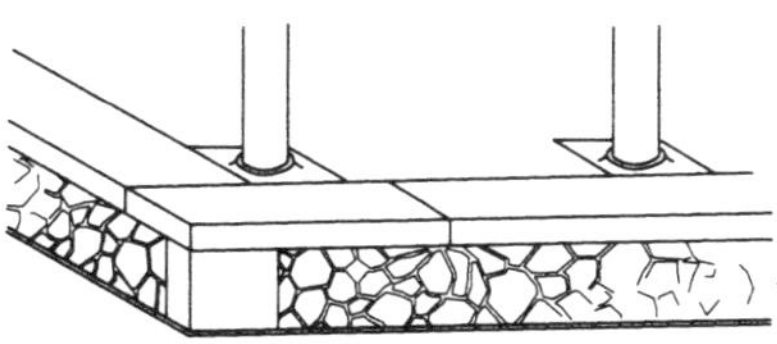

All stone—tiger-skinned stone *taiming*

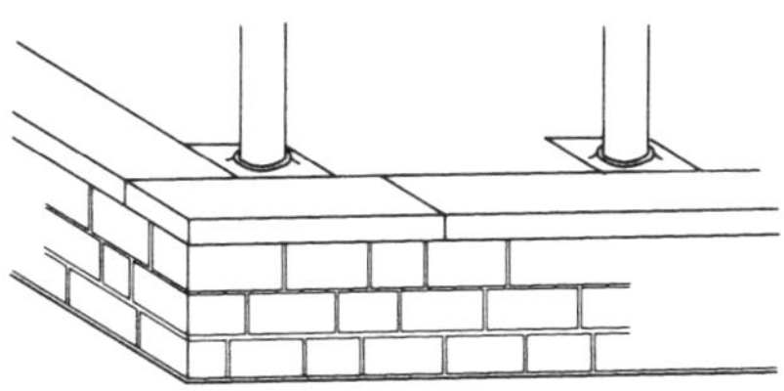

All stone—regular stone *taiming*

Construction ways of ordinary bases.

TOP Brick *taiming.*

BOTTOM *Manzhuang* stone seats.

The soil lining stone is the lowest stone of the exposed part of a base and also the lowest structure that protects the base. The outer edge extends outward from the intermediate piers. The outer edge of the soil-lined stone features *san-shui* (which serves to disperse water). This stone surface is almost flush with the ground or higher on the inside and lower on the outside. For decoration purposes, moldings of the soil-lined stones have been crafted in a variety of shapes, including square openings, round openings, semicircles, and even multi-layers.

In the base plane, a threshold stone of bricks is placed under the threshold. On a platform base of some higher-standard or ceremonial buildings, there are *fenxin* stones (chariot stones) in the middle from the threshold stone to the midline of rectangular stone slabs. *Fenxin* stones are generally long and vertical strips.

In addition to rectangular stone slabs, intermediate piers, and soil lining stones, platform bases were developed with the addition of corner piers around the corner of the base, with pillar stones placed in their middle. Thus, on the left and right sides of a base were corner piers or pillar stones, on the upper and lower sides were rectangular stone slabs and soil-lined stones, and the intermediate piers were sunken in between. All stones were made of carved slates or brick settings. During the prosperous Tang Dynasty, this platform was the standard design. Cornerstones were crafted in various forms, including bas-relief, high relief, and round engravings. However, as platform bases evolved, pillar stones were gradually replaced by intermediate piers, which enhanced strength, while pillar stones became less visible in later periods.

A cornerstone was used as a transition at the corner of the rectangular stone slabs. According to *Yinzao Fashi*, "The rules for making cornerstones: the side length should be two *chi*, for every *chi* increases in side length, the height should be increased by four *cun*. It is not recommended to use corner posts under cornerstones, unless in the case of halls and similar structures." Prior to the Song and Yuan dynasties, cornerstones were carved with images of horned animals. However, this practice ceased after the Ming and Qing dynasties.

Xumizuo ***base***

Another type of platform base is the *xumizuo*. After the Six Dynasties, *xumizuo* began to emerge on platform bases in major buildings, marking another reform in architectural form incurred as a result of the introduction of Buddhism into China. *Xumizuo* was initially employed as the foundation for Buddha statues. *Xumi* is derived from Buddhist scriptures, and the Sanskrit word "sumeru" is transliterated as "Xiu Milu" or "Xumi Tower." The Himalayas, revered as sacred mountains in Buddhism, represent the center of the universe. In this context, *xumizuo* is employed as the base of the Buddha statue to showcase the Buddha's nobility, sovereignty, and supremacy. "Indian architectural forms were introduced into China with the arrival of Buddhism from the East. China chose to select and absorb the most beneficial aspects. Subsequently, other impacts receded, but platforms and all other bases were entirely transformed into the *xumizuo* style."[14]

Xumizuo entered China with Buddha statues, which were gradually used as pagoda bases and later as the foundations of palaces, temples, and other buildings. Its extensive use in the Tang Dynasty elevated it to a prominent status as a form of decoration for platform bases of nearly all objects that required special attention. *Xumizuo* was used for all types of bases and seats, including those for major buildings such as palaces, pagodas, Dhanari columns, and screen walls, as well as for smaller items such as furniture and articles of amusement.

The earliest known example of *xumizuo* was a pagoda base at the sixth cave of the Northern Wei Dynasty in Yungang, Datong.[15] *Xumizuo* is divided into three parts: the upper and lower fillets and fasciae and the *shuyao* (narrow dado) in the middle. There is also *xiaohun*[16] between the fillets and fasciae and the *shuyao*. Finally, there is a symmetrical distribution between fillets and fasciae and *xiaohun*. The I-shaped base of this pagoda, which is wide at the top and bottom but narrow in the middle to form a dado, is the earliest and most basic form of the *xumizuo* in China. Subsequent dynasties all developed and diversified from this basic form.

In the Tang Dynasty, *xumizuo* still had the upper and lower fillets and fasciae. However, in *xumizuo* of some monumental buildings, the upper fillet and fascia, a distinctive feature of the *xumizuo* at that period, vanished. *Shuyao* used posts with *kun*[17] carved in the center, featuring figures, animals or plants. The figures were typically Buddha statues or musical teachers, while the animals were predominantly lions. *Xiaohun* was generally patterned with an overlaying lotus petal in two layers. The ground was covered with a layer of closed lotus petals, on top of which stood a layer of small lotus petals. Meanwhile, the styles of the plain *xiaohun* and of no *xiaohun* also emerged during this period.

The Song Dynasty saw the emergence of more intricate *xumizuo*. Firstly, the upper and lower fillets and fasciae increased to three layers or more. On the upper fillet and fascia, decorative designs were expanded to include patterns of flower and grass, cloud, water, swastika and animals. However, plain fillet and fascia also applied. *Guijiao* was added to the lower fillet and fascia, a design that was consistently used and gradually refined. In addition, *shuyao* became multi-layered with different heights. The main layer of *shuyao* became significantly higher than the other layers. All *shuyao* posts, *kun*, and patterns were inherited from the Tang Dynasty. The number of *hunxiao* increased in line with the number of *shuyao* tiers. The Song Dynasty saw the emergence of multi-tiered *hunxiao*, which typically featured two layers with lotus petal patterns. These were slightly thinner than those seen in the Tang Dynasty.[18] All in all, *xumizuo* in Song was multi-tiered, elegant and tall, brick- or stone-made, and constructed with standards recorded in *Yingzao Fashi* for detailed methods, materials, and sizes.

Since the Yuan Dynasty, *xumizuo* has adopted a simplified form. The majority of these items were restored to their original form, featuring a one-layer *shuyao*, two-layer fillets and fasciae, and two-layer *hunxiao*. The upper fillet and fascia were made very thin, the *shuyao* was lower than in the Tang and Song dynasties, and the *kun* and figure decorations were rare, replaced by ribbon-shaped curly grass. From the Yuan to the Ming to the Qing Dynasty, the *shuyao* remained unchanged.

Xumizuo gained greater popularity in the Ming and Qing dynasties, with its practice illustrated in the *Gongcheng Zuofa Zeli* and *Yingzao Suanli* (*Building Calculation Examples*). The upper and lower fillets and fasciae were of the same height, with continuous patterns such as grass or precious flowers on the upper and simple and elegant decorations on the lower. The *guijiao* below the lower fillet and fascia grew in size and height to accommodate the overall shape. Moreover, *shuyao* employed the use of rosettes, a pattern composing intertwined flowers and plants with streamers at both ends. The corners of *shuyao* were sometimes left unadorned, without any special treatment. At other times, they installed corner pillars called diamond pillars with *ruyi* on both ends. These were known

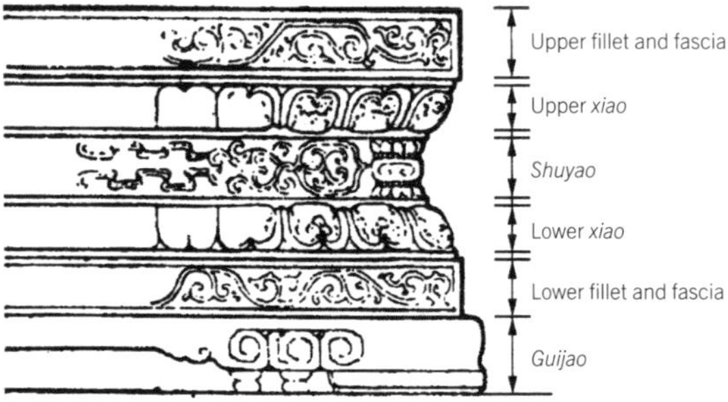

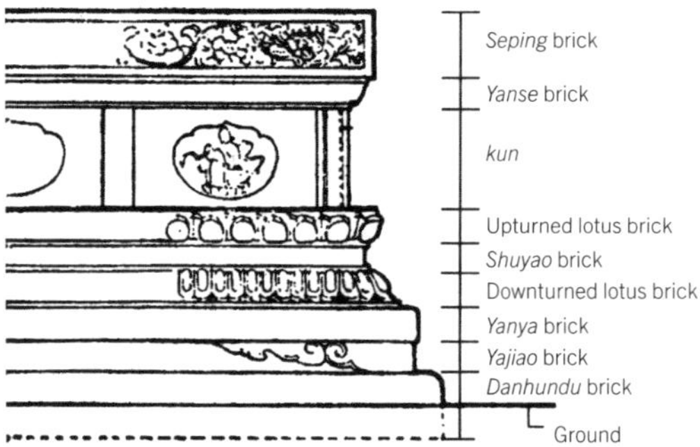

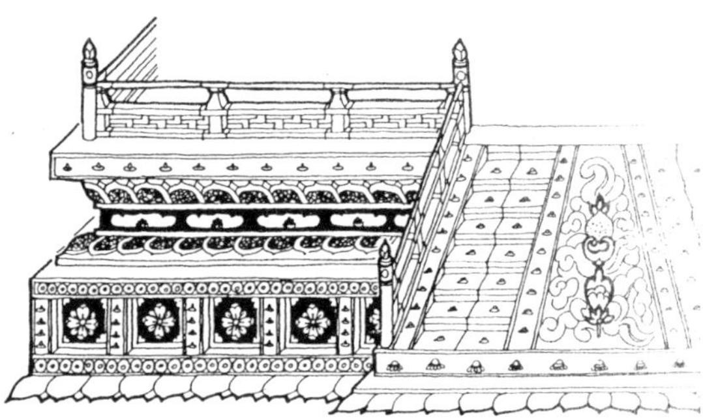

TOP *Xumizuo*, Qing Dynasty.

MIDDLE *Xumizuo*, Song Dynasty.

BOTTOM *Xumizuo* with bi stone, the Middle Tang 231 Grotto mural, Dunhuang.

TOP *Xumizuo* for Pilu Temple Buddha, Longxing Temple, Zhengding County, Hebei Province.

MIDDLE *Xumizuo* base.

BOTTOM *Xumizuo* for Great Compassion Bodhisattva, Dabei Pavilion, Longxing Temple, Zhengding County, Hebei Province.

as *Ruyi* Vajra Pillar. Another style of corner pillar was horseshoe pillar or agate pillar. The upper and lower *hunxiao* were carved with plump and thick lotus petal patterns and sometimes were only with moldings without any patterns. Finally, a faucet named *chishou* (the head of a dragon)[19] was installed on a tall platform base at the corner post on the upper fillet and fascia to disperse water on the terrace. The faucet, which discharges rainwater from its mouth, serves both as a drainage facility and an adornment.

Xumizuo is typically applied to higher-level palaces, temples, Taoist temples, and a few monumental buildings, serving as a structural symbol of medium and high-level platform bases.

In addition to the use of ordinary bases and *xumizuo*, multiple bases, also called composite bases, were employed when enlarging bases for tall and large building volume. This approach aimed to achieve both functional scale and holistic harmony. Given the importance of achieving a high-grade architectural design, multiple *xumizuo* bases became the dominant choice, as they could be used to create an air of solemnity and majesty.

Platform Steps

A platform base has a height, so there is a difference in elevation between the floor and the terrace. This difference is covered by steps, but for low platforms, steps are not necessary. With the exception of rammed earth structures from earlier periods, steps are more commonly constructed from stone, bricks, or masonry. They typically manifest in two forms: layered or sloped.

Layered steps

Layered steps are a basic form of passageway composed of tiers of stacked stone barriers. These techniques were used in half-cave buildings as early as the Neolithic period, with the majority of the material extracted directly from the native soil.[20] Later, rammed earth steps were developed, and in the portrait bricks of the Eastern Han Dynasty, the earliest masonry steps were depicted with drooping belt stones on both sides.

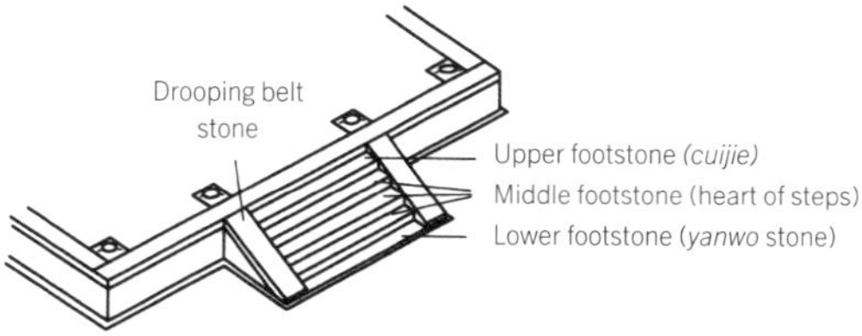

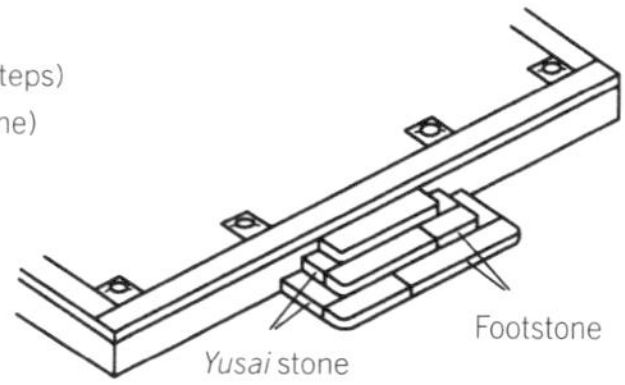

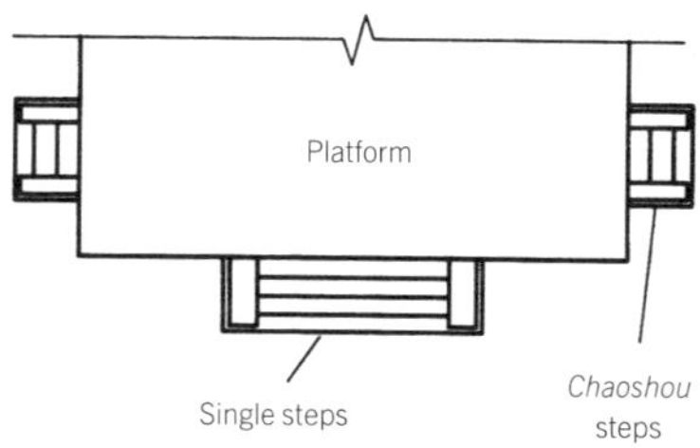

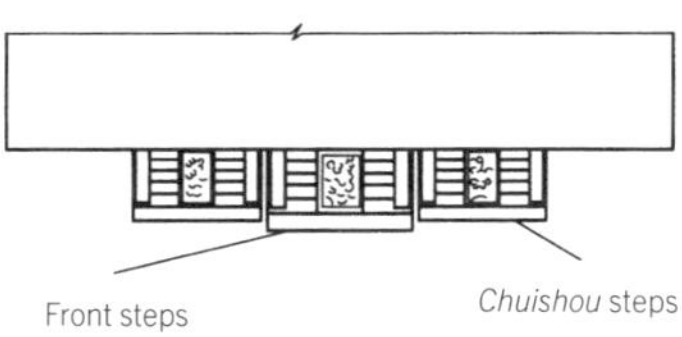

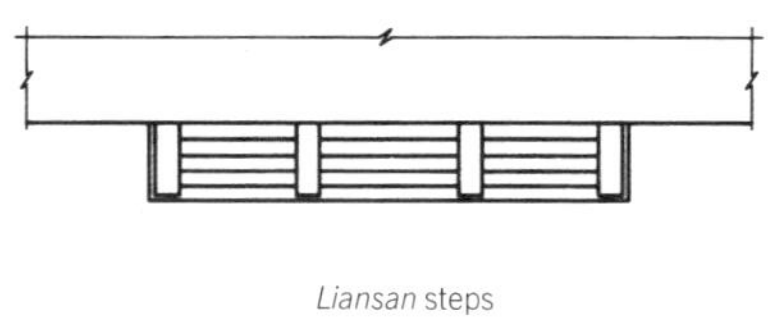

TOP LEFT Drooping belt stone steps.

TOP RIGHT *Ruyi* steps.

MIDDLE LEFT Cloud steps.

MIDDLE RIGHT Single steps and *chaoshou* steps.

BOTTOM LEFT Steps with drooping belt stones.

BOTTOM RIGHT *Liansan* steps.

The structure of layered steps is relatively fixed, with drooping belt, *ruyi* (without drooping belts on both sides of steps), and cloud-step being the most common. According to their applications, steps are classified into the following categories: drooping belts, front steps, single steps, *chaoshou* (placed on the east and west sides of a platform), and *liansan* (three consecutive steps in a row).

TOP LEFT Drooping-belt stone steps.

TOP RIGHT *Ruyi* steps.

BOTTOM LEFT Drooping-belt stone steps, Kaishan Temple, Gaobeidian City, Hebei Province.

BOTTOM RIGHT *Ruyi* steps, Shanliquan Pavilion, Jinci Temple, Shanxi Province.

The drooping belt steps are designed to fix the middle stairs with the side drooping belts. A *yanwo* stone is placed at the lowest step, on which a nest is dug out to rivet and block the flanking drooping belts. Below this, triangular stone walls are constructed. The triangular space stones were originally smooth, but after the Song Dynasty, they became concave on each layer, with only a few of them carved with patterns.

Ruyi steps are defined as stairs without drooping belts. They are designed with a free-style aesthetic and are typically constructed in residential or garden buildings. Some of these stairs have a gradual decrease in length from the bottom to the top. A freer style among *ruyi* steps is the cloud-step, which stacks unworked stones as stairs in their natural shapes to fulfill the basic function of a walkway.

The height-to-width ratio of steps is 1:2, while that of some special shapes is 1:1. The *Yingzao Fashi* outlines the following specifications, "The construction of steps is governed by the following rules: the height is determined by the width of a bay; for one-*chi*-high steps, two stairs are built; each stair is five *cun* high and one *chi* wide, and each drooping belt stone is one *chi* and eight *cun* wide."

In *Mu Jing*, Yu Hao depicted the incline of the steps. There are three levels: steep, flat, and slow. The palace provides a suitable reference point for the imperial chariot. When the rear charioteers begin to climb from the bottom step, the front charioteers hang down their arms holding chariot poles, and the rear people elevate the poles to their arm's utmost; this is the steep level. When the front poles are flush with the charioteers' elbows, and the rear poles are right on the charioteers' shoulders, this is the slow path. When the front poles hang down and the rear poles are flat on charioteers' shoulders, this is the flat level. By raising the arms of the front and rear charioteers, Yu Hao clarified the different slopes of the steps. Based on the specifications for the human body in the *Huangdi Neijing Taisu*: the upper body and legs are 6.2 feet, the upper arms 1.7 feet, and the lower arms 1.65 feet, Joseph Needham calculated the slope ratios for the three levels depicted in *Mu Jing* as follows: 3.35 for steep, 1.38 for slow, and 2.18 for flat. These proportions align with the lower portion of Yu Hao's statement, "Every house has three parts." It should be noted that the slope ratios may vary in practical step construction.

Steep, flat, and slow levels in *Mu Jing*.

Sloped steps

The use of sloped steps eliminates the need for additional layers. The slope from the floor to the terrace is primarily designed to facilitate the passage of vehicles and horses. This type of step employs bricks and stones to form a flat slope. The gaps or rough surfaces between bricks and stones serve to produce friction, thereby ensuring a smooth passage and an anti-slip function. However, over time, the friction between the bricks and stones will diminish, making the passageway slippery.

To enhance their anti-slip properties, sloped steps are constructed with masonry side edges to create a jagged slope surface known as *jiangcha*. In *Yingzao Fashi*, *jiangcha* is referred to as a slow road: "The regulations for building a slow road are as follows: for city gates, the height-slope ratio is 1:5 (the width of a slow road one foot shorter than the terrace); for halls, the ratio is 1:4 in a style of three-cicada-wing (the middle level is the same as that of the bay, and the width of two sides is 4 *cun* per *chi* of diagonal length, all varying according to the specific situation). The width of the middle level is the same as that of the bay, and the width of two sides is 4 *cun* per *chi* of diagonal length, with variations permitted according to specific circumstances. The side slope and the middle

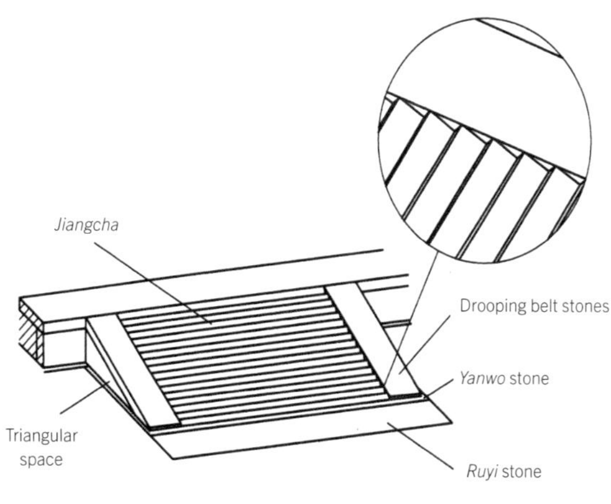

Jiangcha.

level are constructed in accordance with step construction. For a five-cicada-wing style (the middle level is flat and four sides are beveled), the width of four sides is 3/4 of the diagonal length. All face bricks are arranged to create a sloped surface with a depth of approximately 10 mm, extending approximately three-thirds of the brick's length. If the bricks are not unsmooth, then a jagged slope is unnecessary." *Jiangcha* provides superior anti-slip properties and can also be used as pedestrian ramps.

The highest level of steps is called *danbi* and is found in imperial palaces. The term "bi" is defined in Liu Shen's *Shuowen*, "It refers to steps up to a high place, pronounced as *bi*." In essence, the original meaning of *bi* as "steps" was transformed over time to refer exclusively to the steps leading up to an emperor's palace. In ancient times, the terms *jie* and *bi* were used interchangeably. Since the Qin and Han dynasties, *bi* has been the specific name for the steps of palaces, altars, and temples.[21] This explanation has its roots in a historical context. Cai Yong's *Du Duan*, Volume 1 stated, "When ministers present reports to the emperor, they do not address the emperor directly but instead call the attendants at the bottom of the steps to offer the reports. This reflects the respect that subjects have for the emperor." Over time, *bixia* (the bottom of steps under the throne) became the honorific title for emperors. This was because attendants stood below the emperor's throne in the imperial palace, and *bixia* could be referred to as "under the steps of the emperor's palace." This explanation seems to make sense. In addition to the emperor's wife and concubines, his sons and daughters, and his brothers and sisters, there was another title for them all: *dianxia* (literal translation: under the palace). This "palace" was considered to be less impressive than the emperor's palace, but in fact the emperor also slept in his wife and concubine's palaces. The usage of *dianxia* seems to be confused with *bixia* in terms of the hierarchy of palaces. In fact, Liu Shen defined *bi* as steps to climb high, but these steps were reserved exclusively for the emperor. The steps then evolved from a high platform between the Warring States Period and the Qin and Han dynasties, where the emperor ascended as the Son of Heaven to reconcile with Heaven, to steps below the emperor's throne. Therefore, *bi* in the early days was used to refer to steps used by emperors to climb high during the period of high platform construction. It was later specifically referred to steps under the emperor's throne, and then was extended to cover steps of all palaces used by the emperor. This evolution of bi clarifies the hierarchical relationship between *bixia* and *dianxia*.

The steps in palaces may be framed as steps on both sides and a stone slope carved with dragons, phoenixes, and cirrus clouds in the middle. The middle slope is known as the *yu* (imperial) road or *bi* stone. This architectural element first appeared on the platform under the seat of the emperor or Buddha in earlier times. It later became a feature of important palaces, temples, and other buildings. These platforms were carved with white or plain marble, featuring a combination of bas-relief and line carving. They were adorned with dragons, phoenixes, cloud patterns, water ripples, group flowers, and *baoxiang* flowers. Li Yunhe posited that this structure, comprising two steps and one road, may be a vestige of an ancient two-level system. This system merged the east and west steps, creating a royal road that resembled a red carpet in the west, reserved for the most distinguished individuals.[22] This conclusion is not supported by the practical function of the *yu* road. The slanted bas-relief *yu* road is uneven, unstable, and unwalkable. In fact, Mr. Joseph Needham previously stated that the *yu* road should be a "spiritual road full of reliefs," which is more ceremonial than functional.

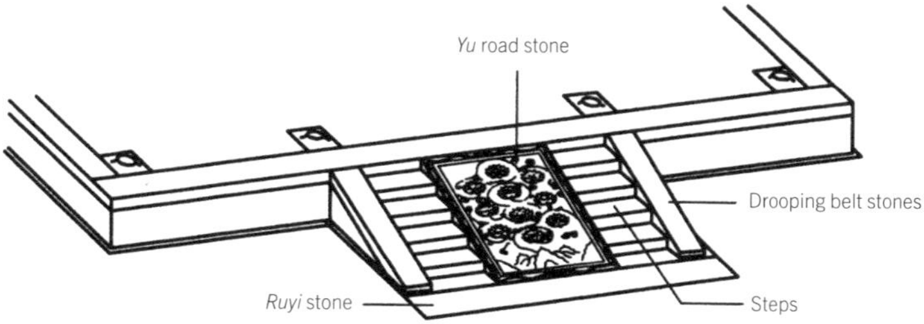

Yu road steps.

Yu road stone, Hall of Preserving Harmony, Forbidden City.

LEFT Dragon pattern of *yu* road stone.

RIGHT Baoxiang flower pattern of *yu* road stone.

Railings

Railings are constructed in conjunction with a platform base, with the primary objective of enclosing the space to prevent people from advancing or falling. It is typical for the railings and platform base to be integrated. As with platform bases, which are an essential component of Chinese building facades, railings are also an indispensable feature. In Chinese architecture, equal attention is paid to the organization and segmentation of space and to function and visual form. This tendency influences the railings' compositions.

In ancient times, a railing was referred to as *langan*, which translates to "crisscrossing." The vertical wood *lan* and horizontal wood *gan* constituted a railing, with the crisscrossing woods forming the framework. In the early days, railings were typically constructed from wood. However, as wooden railings, particularly those used outdoors, were prone to rot due to exposure to wind and rain, the stone began to be used as a substitute for its strength and durability. Stone railings, typically situated outdoors, were utilized in conjunction with wooden ones, which were tailored to specific locations. Stone railings typically emulate the shapes and structures of their wooden counterparts, mirroring the imitation of wooden equivalents in towers or other Chinese architecture seen with stone, brick, and iron *dougong*s. Later, brick or iron railings also emerged, essentially replicating the shapes and structures of wooden railings. However, due to their limited range of use, their structures were not systematized.

A wooden straight railing dating back to more than 7,000 years ago was discovered at the Neolithic settlement site of Hemudu in Yuyao, Zhejiang.[23] The railing was composed of vertical and horizontal wood elements. The vertical wood *lan* was later called a "baluster," while the upper part a crossbar used as the handrail to be *gan*, named *xunzhang* (handrail). The lower crossbar was called sill. A railing was constructed by extending a rectangular frame constituted of baluster, *xunzhang*, and sill in a continuous manner. This is the genesis of railings. All subsequent railings can be traced back to this original production, with the exception of variations in decoration.

The greatest variety in railing embellishments occurs in the structures, designs, and patterns of the basic units of the rectangular frame. The frame is treated virtually or authentically for functional and visual considerations. In terms of functionality, builders may select structures that either prevent or allow wind to pass through or that convey a sense of stability or speed. Some frame units are processed to be completely transparent, some to be semi-transparent, and others to be impermeable. However, transparent *xunzhang* and solid frieze panels are more often employed. Given that railings are designed to be visually prominent, ancient craftsmen observed the law of unity and continuity to ensure a harmonious connection between doors, windows, and railings.

The Han Dynasty is notable for its use of simple and elegant railings in a number of its paintings and pottery housewares. These designs were developed

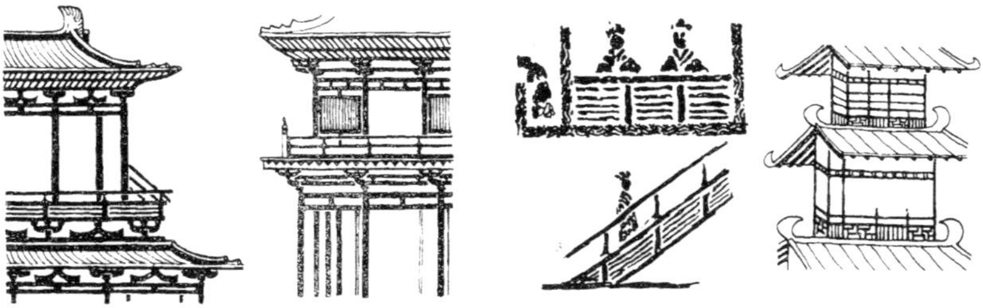

Images of railings in reliefs, murals, and stone engravings.

with great care, combining a baluster, *xunzhang*, and frieze panel in a sophisticated composition. The frieze panels exhibited a variety of forms, including straight bars, horizontal bars, oblique bars, and intersected circles.

As late as the Sui Dynasty, *penchun* (lip) and king posts were still being used in railings. Below the *xunzhang*, *penchun* was installed above the frieze panel. In the middle, a king post supported *penchun* and *xunzhang*, making the *xunzhang* more sturdy and durable and the railing frame richer and airier.

In the Tang Dynasty, railings incorporated the fundamental techniques of preceding dynasties while showcasing a more diverse and visually appealing aesthetic. The use of ornamental patterns was extensive, and they are applied to *xunzhang*s, frieze panels, and balusters. The frieze panels were available in two types: solid and open-carved. Additionally, balusters and king posts were enhanced. The overall presentation was elegant and luxurious.

Up until the Tang Dynasty, *xunzhang*s on wooden railings were typically designed with a long, uninterrupted length, with corners handled in two distinct ways: "*xunzhang*s crossed at right angles around the corner" and "*xunzhang*s met at right angles around the corner." The first treatment did not require a baluster around the corner, with two *xunzhang*s intersecting at a 90-degree angle. In contrast, the other way required a baluster for the two *xunzhang*s to meet at a 90-degree angle around the corner. In addition to king posts, *cuoxiang* (concave-shaped) was incorporated to reinforce *xunzhang*s. Balusters were available in a variety of shapes, including square, circle, octagon, and melon prism.

Yingzao Fashi in the Song Dynasty recorded two types of stone railings: single and double frieze balustrades. From a structural standpoint, they remained part of timber structures, and neither *xunzhang* nor *penchun* was suitable for stone processing.

There were discrepancies between single and double frieze balustrades in the Song Dynasty. Firstly, double frieze balustrades were perceived to be of a higher rank and larger in size than single ones. The height of the double frieze balustrades was 4 *chi*, with each section measuring 7 *chi* in length. In contrast, the single frieze balustrades were 3.5 *chi* in height and each section was 6 *chi* long. Furthermore, double frieze balustrades comprised two layers of frieze panels. The supports of *xunzhang* above *penchun* combined *yungong* (shaped like clouds) and *yingxiang* (bulging part in a king post). The baluster (octagonal in cross section, with a head usually shaped like a lion sitting on upturned and downturned lotuses) was embellished with patterns and its column base with panned lotus. Single frieze balustrades, on the other hand, only had one layer of Svastika frieze panels. Below *xunzhang*, a combination of thinner *yungong* and *cuoxiang* was used. Balusters were made of plain stones.

From top to bottom, components of a baluster for a double frieze balustrade were *xunzhang*, *yungong*, *yingxiang*, *penchun*, big frieze panel (with king posts), *shuyao*, and small frieze panel (without king posts, also called *dixia*), and sill; for a single frieze balustrade were *xunzhang*, *yungong*, *cuoxiang*, *penchun*, Svastika frieze panel and sill. Not every section of the Song-style balustrade ended by balusters, which were more often employed around the corners and at the beginning; this is an expression of characteristics of wooden frames.[24] Sometimes, balustrades started and ended with drum-shaped bearing stones, and this processing method influenced later eras.

The railings in the Qing Dynasty underwent significant changes, with a more rational structure, a more compact shape, and a notable shift in the overall visual impact. The cross-section of the *xunzhang*, which previously had a round shape in the Song Dynasty, was redesigned as a polygonal square that was stronger and more conducive to stone processing. Below the *xunzhang* stood a lotus-leaf "purity vase" (in Sanskrit: kalasa), and between the two balusters was one "purity vase" in the middle and half of a "purity vase" placed at each end of a baluster. The panels were typically composed of a single layer of unadorned stone, with only a few carved patterns and undulations.

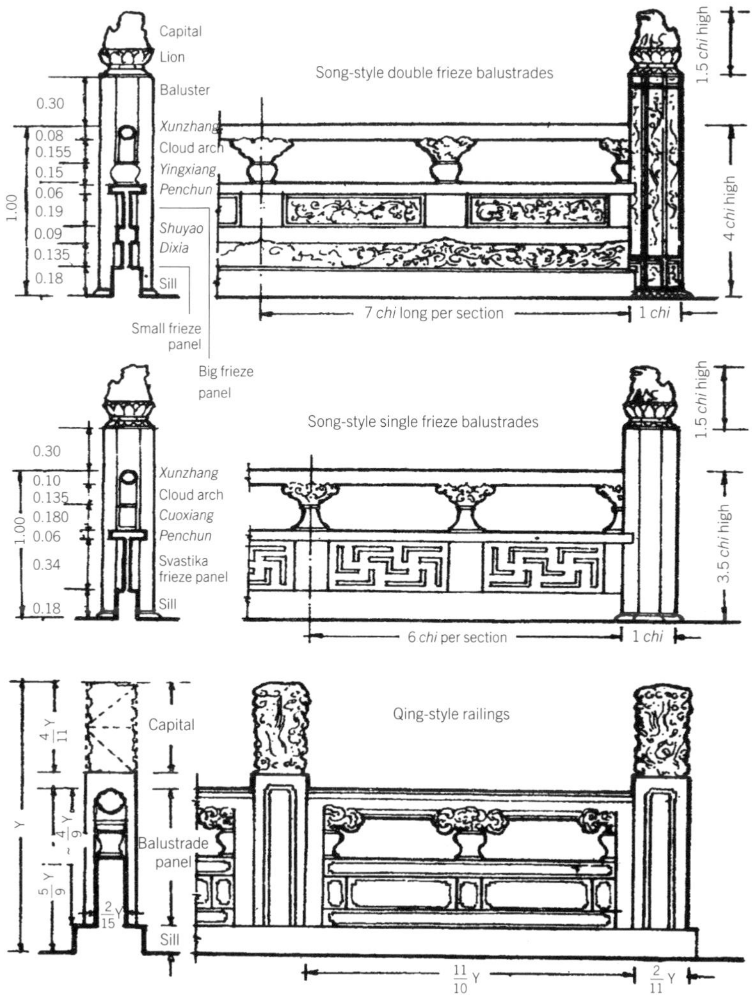

Song-style double and single frieze balustrades and Qing-style railings.

Lotus capital—downturned lotus capital, mostly used in gardens

Banana leaf capital, mostly used in palace gardens

Lotus capital—overturned lotus capital, mostly used in gardens

Water pattern capital, mostly used in gardens or specific environments

Dragon capital, mostly used in palace gardens

Horsehead capital, commonly applied or locally practiced

Styles of capitals.

The Qing-style railings emphasized the use of balusters in each standard section. Consequently, the balusters underwent the most significant transformation in the entire vision. In contrast to the Song style, balusters in the Qing style were erected directly on the sill through joggle joints, eliminating the need for a column base. In addition, they became square columns with simplified shafts and enriched capitals that were enlarged and heightened. A variety of stylized patterns were applied to baluster capitals, including cloud dragon, cloud phoenix, flame, stacked cloud, pomegranate, overturned lotus, lion, and plain square.

The size of the Qing railings was not fixed, which allowed balusters to serve as a reference for other construction dimensions. The *Yingzao Fashi* of the Qing Dynasty set forth the following specifications: "For railings installed on the terrace of the base in a palace, the height of balusters is 19/20s of the base height, which is 1/4 height between the curb and wood-plate lintel. The square section of balusters is 2/11s of the base height, and the lower tenon is 3/10s of the baluster square. Finally, the height of the column capitals is double the baluster square."

Depending on the location and adaptability with other structures, railings fulfilled multiple functions, including as bench railings or backrest railings in pavilions and colonnades, and wooden railings in furniture.

To prevent water from overflowing the platform base, a drainage outlet called *chishou* (the head of a dragon) extended from the platform base under the balusters. This outlet was designed to drain accumulated water. The Chinese character for "chi," which is the sixth son of the dragon, represents the ability to swallow rivers and spit out water. While it was highly ornamental, it also served as a symbol to pray for blessings and good luck.

TOP Railings, Faxing Temple, Zhangzi County, Shanxi Province.

BOTTOM Qing-style railings, Summer Palace, Beijing.

Pavage

In ancient times, indoor floors were typically constructed using rammed earth,[25] a technique that is still employed in many modern villages and less important rooms of houses in remote villages. Subsequently, lime-mixed soil floors and brick-paved floors gradually became more prevalent, while wooden strips were used to create floors in some important houses. Lime-mixed floors were more prevalent in the southern regions, while brick and stone paving floors were more common in the northern regions. These applications remained in use for several centuries in Chinese history.

In the ruins of the primitive society, it was discovered that the ground had been hardened by burning to level it and prevent moisture. A large central square with an area of 1,300 m^2 was discovered at the late Neolithic cultural site of Yuchi Temple in Mengcheng, Anhui Province. The circular structure was paved with a 10 cm thick layer of red-burnt soil particles, creating a smooth surface. From the cross-section, it was evident that the area had been artificially bedded. In the ruins of the middle Neolithic period, outdoor pebble pavements were introduced. The practice of laying pebbles vertically for water distribution began at least from the middle of the Western Zhou Dynasty and was largely replaced by bricks after the Tang Dynasty.

The earliest known examples of floor tiles were discovered in the Qijia Village area of Fufeng, Shaanxi Province, around the late Western Zhou Dynasty. The tiles were 50 cm square with round protrusions at the four corners at the bottom for fixation with the cushion.

In the ruins of the ancient city of Zhenghan in Xinzheng, Henan, floor tiles with corner edges on all sides were excavated during the Spring and Autumn and Warring States periods. The tiles were slightly smaller, measuring approximately 35 to 45 cm^2. In addition to plain tiles, the surface was adorned with Chinese character 米, twisted rope, and fret patterns to prevent slipping.

Bricks with sawtooth cross-sections and mortises and tenons emerged in the remains of the Qin Dynasty. The brick dimensions were 50 cm × 35 cm × 5 cm. In the pit of Qin Shihuang's Terracotta Warriors and Horses, there were slightly wedge-shaped paving tiles.[26] After the Han Dynasty, strip bricks and square tiles were popular. In Han tombs, various forms of pavings were used, including stone slabs and hollow-brick pavements. During the Eastern Han Dynasty, enhanced techniques were employed to grind bricks to create flush seams. In the Tang Dynasty, the sides of floor tiles were ground into an inclined plane, resulting in virtually invisible seams between bricks when connected on the surface. This bondage between bricks became stronger and more reliable. Lime bonding gained significant traction in the Song Dynasty, significantly enhancing the waterproofing and adhesion of brick paving.

Brick paving is an effective solution for both anti-slip and decorative purposes. The floor tiles are patterned on the surface and in paving methods. The surface patterns include fret from the Qin Dynasty, four-spirit from the Han Dynasty, and combinations of words and patterns, such as auspicious text such as endless joys for all ages, and also pearl lotus patterns in the Tang Dynasty.

The brick paving patterns vary considerably depending on locations. In general, palace buildings employ steep slab cross seams, square brick sloping weirs, oblique willow leaves, and straight willow leaves. Small buildings utilize square brick cross seams, strip brick cross seams, *guaizi* brocade, and strip brick slope. In residential and garden buildings, we offer a range of patterns, including *bafang* brocade, sets of *bafang*, mat, herringbone, and willow leaf herringbone. We also provide auspicious textures such as turtle-back brocade, trigram brocade, and Svastika brocade. In the residences and gardens of the Ming and Qing dynasties, a variety of materials were used in combination, and construction wastes such as broken bricks, gravel, broken tiles, ceramic tiles, and pebbles[27] were repurposed into geometric designs, animals, plants, and text patterns. The forms and graphics are diverse and adaptable to local conditions and materials. In regions south of the Yangtze River, this paving method is known as "Flower Street Paving," which is largely applied to outdoor and courtyard floor paving.

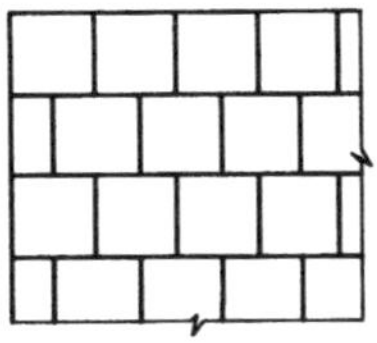

Square brick cross seams, common style

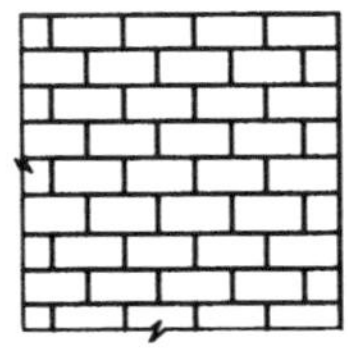

Strip brick cross seams, mostly used in (the) inside and outside of small buildings

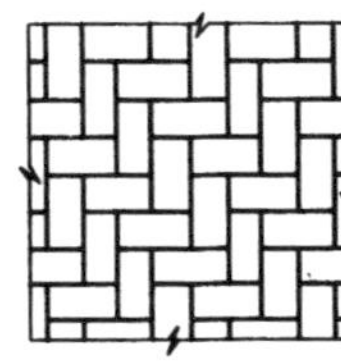

Guaizi brocade, mostly used in (the) inside and outside of small buildings

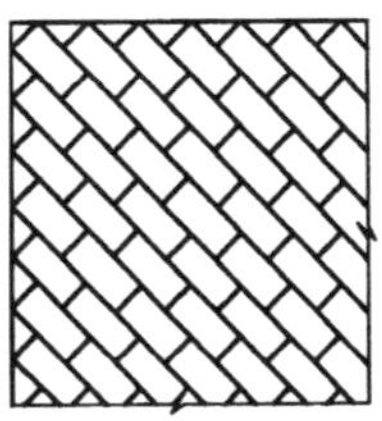

Strip brick sloping weirs, mostly used in small buildings

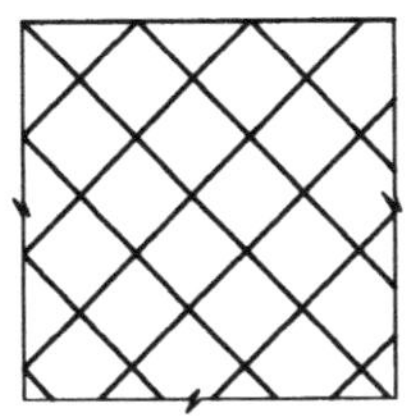

Square brick sloping weirs, mostly used in superior buildings

City wall brick steep slab cross seams, mostly used in palace buildings

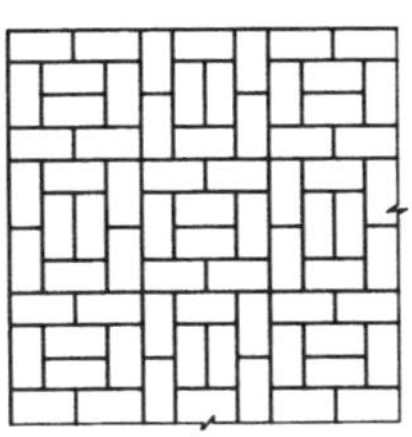

Sets of *bafang*, mostly used in residences or gardens

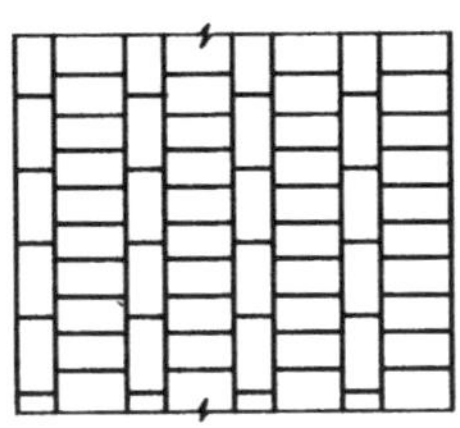

One straight and one horizontal, mostly used in gardens

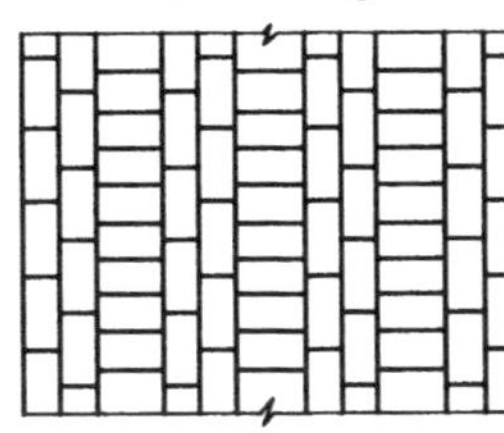

Two straights and one horizontal, mostly used in gardens

Ways to pave floor bricks.

NOTES

1. The *Mu Jing* is believed to have been authored by Yu Hao, a native of eastern Zhejiang at the end of the Five Dynasties and the beginning of the Northern Song Dynasty. Yu Hao was a construction craftsman from a humble background. The dates of his birth and death are unknown. Due to the lack of historical records, we only know that he worked as a craftsman (carpenter in charge of design and construction) in the early Northern Song Dynasty and was engaged in architectural practice for a long time. In the Song Dynasty, Ouyang Xiu's *Guitian Record Volume 1* records that Yu Hao designed and built the Kaifeng Kaibao Temple Tower, a wooden tower tilted slightly to the northwest. He predicted that the tower would be blown upright by the wind within a hundred years. Ouyang lauded him as "the most accomplished carpenter since the founding of the Song Empire." It is said that Yu Hao authored three volumes of the *Mu Jing* in his later years. The *Mu Jing* is a work on house construction methods, but was later lost to the world. There are brief references to this work in Shen Kuo's *Mengxi Bitan.*
2. Liang Sicheng, *The Complete Works of Liang Sicheng*, vol. 6 (Beijing: China Construction Industry Press, 2001), 237.
3. Li Yunhe, *Principles of Chinese Classical Architectural Design* (Tianjin: Tianjin University Press, 2005), 12.
4. The walls and spaces under the eaves of Chinese buildings form a secondary space that transitions from outdoor space to indoor space. In modern architecture, this is referred to as the gray space. This space is also a harmonious transition of the overall vision.
5. Liang Sicheng, *Examples of Qing-Style Construction* (Beijing: Tsinghua University Press, 2006), 45.
6. The Chengtoushan ancient cultural site is located in Li County, Changde City, Hunan Province. The site encompasses a period from the prehistoric Daxi culture to the Shijiahe culture period in southern China. It is also the only ancient city site in China with the earliest discovery and the most complete protection to date.
7. Hunan Provincial Institute of Cultural Relics and Archaeology, *Chengtoushan, Li County: Excavation Report of Neolithic Age Sites (Part 1)* (Beijing: Cultural Relics Press, 2007), 89.
8. The Erlitou site is located in Yanshi City in the eastern part of the Luoyang Basin. It has been initially confirmed to be the capital site of the middle and late Xia Dynasty, with an estimated age of 1750 BC to 1500 BC. The Erlitou site offers invaluable insights into the origins of Chinese civilization, the rise of countries, the origin of cities, the construction of royal capitals, the customization of royal palaces, and other major issues. It is recognized by the academic community as one of the most significant ancient cultural sites in China.

9. The Shangcheng Ruins in Zhengzhou are capital ruins from the mid-Shang Dynasty. It is located in Guancheng District, Zhengzhou, with a total area of 25 square kilometers. It is a significant capital ruin second only to the Yin Ruins in the pre-Zhou Dynasty. The city wall was first built in 1500 BC. Some scholars believe that this is the capital of Bo where "Tang first lived in Bo."
10. Li Yunhe, *Principles of Chinese Classical Architectural Design* (Tianjin: Tianjin University Press, 2005), 173.
11. *Mingtang*, *shishi*, and *chongwu* are locations where ancient emperors "articulated their political and religious ideologies." The site was known as *shishi* during the Xia Dynasty, *chongwu* during the Shang Dynasty, and *mingtang* during the Zhou Dynasty. All major ceremonies, including court gatherings, sacrifices, celebrations, the selection of scholars, elderly care, and teaching, were held there. From ancient times to the Tang and Song dynasties, with different systems in place, the shapes and scales of construction in each dynasty varied. This has led to differing interpretations from academic circles.
12. Liu Jie, *Chinese Traditional Architectural Decorative Art: Platform* (Beijing: China Architecture Industry Press, 2010), 24.
13. In the Song Dynasty's *Yingzao Fashi*, it was called *yalan* stone; in the Qing Dynasty's *Gongcheng Zuofa Zeli*, it was called *jietiao* stone; and in the Wu Dynasty's *Yingzao Fayuan* (*Building Origin*), it was called *jieyan* stone.
14. Li Yunhe, *Principles of Chinese Classical Architectural Design* (Tianjin: Tianjin University Press, 2005), 175.
15. Pan Guxi, *The History of Chinese Architecture* (Beijing: China Architecture & Building Press, 2015), 264.
16. Refers to the convex and concave decorative line corners; its cross section is called *xiao* for the upper half in the shape of a concave arc, *hun* (or half *hun*) for the lower half in the shape of an outer convex arc; it can also be evenly divided into three parts, the top is *xiao*, the middle is the furnace mouth, and the bottom is *hun*. In some cases, the *yin* and *yang* bricks found in ancient buildings are also referred to as *xiaohun*, which are composed of *xiao* with concave arc-shaped sides and *hun* with convex arc-shaped sides.
17. The name "kun" is written as "humen" in *Yingshi Fazao*. It is suspected that this was an error in multiple editions and should have been written as "hu." Zhang Yuhuan believes that "kun" is actually a Buddhist niche commonly used in Buddhism. *Kun* appeared when the image of the niche was removed and carved with lines. The outline of *kun* can be arc-shaped, rectangular, oblong, etc., with a protrusion in the middle of the upper end, shaped like a gourd spout. Entrances in Buddhist buildings that can show dignity usually use the *kun* style, such as Buddhist tents, niches, and halls. The interiors are often carved with Buddha statues, Bodhisattvas, Buddhist-themed stories, and patterns of plants and animals.

18. Li Jinlong, *Identifying Chinese and Foreign Ancient Buildings* (Shanghai: Shanghai Bookstore Publishing House, 2016), 16.
19. According to ancient legend, *chi* is a dragon with no horns. The Chinese dragon has six sons, the sixth of whom is *chi*. It has the ability to consume vast quantities of water and then expel it. It is a highly decorative item with a symbolic meaning of praying for blessings and bringing good luck.
20. Pan Guxi, *The History of Chinese Architecture* (Beijing: China Architecture & Building Press, 2015), 264.
21. Li Yunhe, *Principles of Chinese Classical Architectural Design* (Tianjin: Tianjin University Press, 2005), 313.
22. Ibid., 177.
23. Pan Guxi, *The History of Chinese Architecture* (Beijing: China Architecture & Building Press, 2015), 265.
24. Li Yunhe, *Principles of Chinese Classical Architectural Design* (Tianjin: Tianjin University Press, 2005), 254.
25. Joseph Needham, "Civil Engineering and Nautical Technology," in *Science and Civilisation in China*, vol. 4 (Beijing: Science Press, 2008), 135.
26. Pan Guxi, *The History of Chinese Architecture* (Beijing: China Architecture & Building Press, 2015), 266.
27. This paving method was developed to utilize construction leftover materials and address the limited nature of construction waste. Proprietary materials were gradually developed to meet this need. This method is environmentally friendly, economical, and free to create, which has led to its widespread use in subsequent courtyard paving.

CHAPTER 4

TIMBER FRAME

Structural Characteristics of Timber Frames

Chinese architecture, from the earliest building ruins for study—the above-ground and semi-cave-dwelling houses of the primitive society in Banpo, Xi'an—evolved and developed intricately over tens of millions of years. Building structures are technically developed in two directions: load-bearing walls or load-bearing frames. From ancient dynasties to later periods, these two structures appeared simultaneously or cooperatively. Meanwhile, timber frames replacing stone structures became a mainstream of Chinese buildings. Whether the early rammed earth or the adobe brick walls after the Qin and Han dynasties, both were intended to fulfill the functions of enclosing a house and supporting loads, but neither was better than a timber frame that could easily support weight and be enclosed by small amounts of earth or brick.

A timber-frame house employs wooden roof-trusses or frames in manifold forms as the load body for an entire building. The roof-trusses are composed of columns, beams, and *dougong*s. The roof weight is transmitted through layers of beams from the *dougong* to the columns. Meanwhile, walls and partitions do not carry weight but divide and enclose a house.

Characteristics of timber frames

The use of timber as a frame offers distinct benefits.

First and foremost, wood is a readily available material. In earlier China, under the self-sufficient natural economy, those outside mountainous areas had to purchase and pay for stone processing if they chose stones to build a house. Timber was also a convenient building material, as it could be planted as trees before and behind the house, along roads and rivers, in fields, and in areas where large areas of cereals failed to be grown. It typically takes between three and five years for a tree to reach the point where it can be harvested for timber, although it can take up to ten years. Today, farmers in the vicinity of Xi'an, Shaanxi Province, refer to trees planted by them as pillars, beams, or rafters1.[1] Once they had accumulated sufficient wood, they were able to obtain building materials without having to make significant financial outlays.

Furthermore, wood is readily available in both light and tenacious varieties, making it an ideal material for carving. Timber frame construction is also straightforward. The structure was highly patterned, making it straightforward for local carpenters with the requisite skills to construct. With the guidance of several master carpenters and the division of labor among helpers, a residential house could be completed in a relatively short time.

Additionally, timber-frame houses are highly adaptable to varying terrain. Even in narrow mountainous areas, the lengths of different pillars can be adjusted to suit the terrain, and the construction cost is low since the mountain shape requires minimal leveling.

Finally, the expansion of timber structures is a convenient and efficient operation. The wooden-frame building has a rectangular footprint with bays as units. Should a house require expansion, simply add bays to the original house to increase its surface area. Furthermore, there are numerous ways to expand a house. Similarly, the dismantling and rebuilding of wooden buildings to another location is a straightforward process, with minimal material loss. The Eastern Wei Dynasty is a case in point, as it demolished the Luoyang Palace to Yecheng.

However, it should be noted that the wooden structure also has certain inherent weaknesses. For instance, while wooden structures are effective in vertical load-bearing, they are less robust in horizontal thrust. This is due to the difficulty in creating a tightly joined joint between the timber frames and their foundation. Consequently, when strong winds are blowing, horizontal forces become so enormous that houses are prone to collapse. This creates a need to build thicker enclosures (in addition to preventing cold and keeping warm).

Classifications of timber frames

The classification of timber frame is based on its type, scale, and form. The *Yingzao Fashi* document identifies three types: palace structure, hall structure, and cluster angle beam structure. The *Gongcheng Zuofa Zeli* of the Qing Dynasty describes the scale of buildings into timber frames with *dougong* or without *dougong*. Another division is based on the type of roof, including hip, overhanging gable, or flush gable. Regardless of the classification, timber frames are constructed in three ways: post and lintel, column and tie, and log cabin.

Post and lintel construction. This style was widely used in Chinese architecture, taking shape around the late Spring and Autumn Period. The structure consisted of four main parts: the upright load-bearing columns at the bottom, the load-bearing beams, purlins, and rafters at the top, and in the middle, the beams and *dougong* used to connect and distribute the weight. The construction process was as follows: In the width direction of the house, columns were set up, beams were placed on the columns, and short posts were installed on the beams. This process was repeated until the ridge was reached, where the ridge purlin was placed. The purlins were installed over the beam heads as well as over the short posts above the beams. The rafters were then installed vertically, with the weight of the roof transmitted from the rafters to the purlins, then to the beams, and finally to the columns. Tiebeams were used as bonds among the beam frame. Two sets of beams formed a bay, with the number of bays increasing as the number of beams built did as well. Post and lintel construction took various forms due to different construction methods across eras and regions.

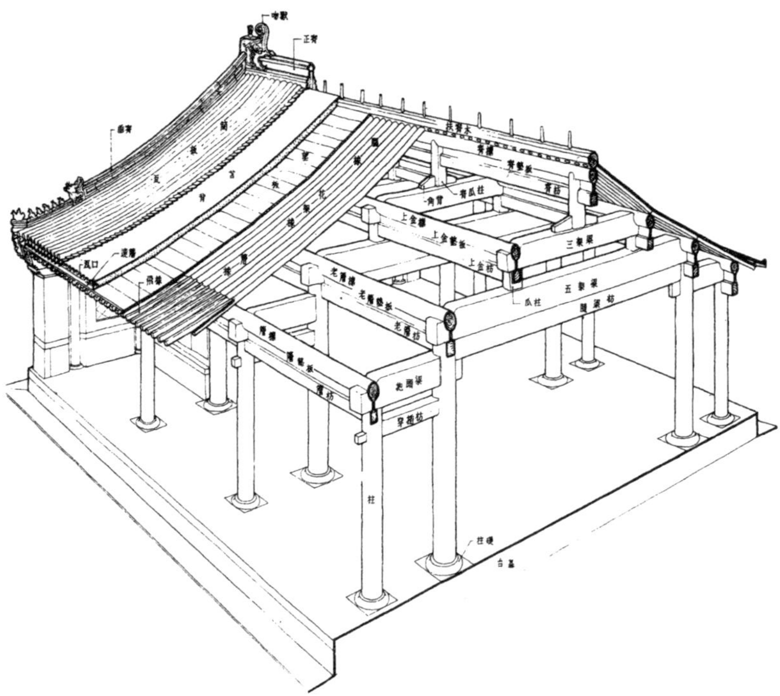

Timber building of post and lintel construction.

Buildings constructed in this manner were typically found in areas north of the Yangtze River, which featured complex structures, abundant materials, spacious interiors, and strong, durable, and imposing buildings. To apply post and lintel construction to small-scale houses, builders removed *dougong* by resting beams and purlins directly on columns, which was called column-beam work. This was classified as a timber frame without *dougong* in the Qing Dynasty and emerged more on overhanging gable roofs.[2]

This category could be further divided into the palace and hall structures recorded in *Yingzao Fashi* of the Song Dynasty.

The palace structure was composed of three horizontal layers superposed from bottom to up: column layer, *dougong* (including *dougong*s, column-top ties, and exposed beams) layers, and roof truss layer. The interior and exterior columns were of the same height. The *dougong* layers constructed a horizontal grid placed on the column grid in order to maintain the stability of the whole frame, which was equivalent to the ring beams in modern architecture.[3]

Timber building, the palace structure with inner and outer columns of equal height in post and lintel construction.

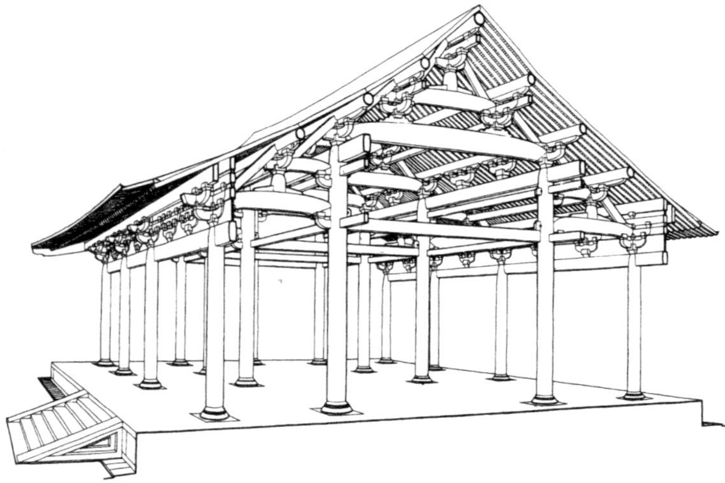

Timber building, the hall structure with inner and outer columns of unequal heights in post and lintel construction.

The hall structure comprised a set of relatively independent vertical frames employed in a variety of ways in one building. Columns were diversified in number, position and height due to construction needs. The palace style can be described as a horizontal-layer superposition, while the hall style is a combination of vertical frames.

The earliest example of a palace structure is the main hall of Foguang Temple, while the main hall of Nanchan Temple belongs to the hall style.

The palace-style structure underwent changes when it was passed on to the Ming and Qing Dynasties. In some high-ranking palaces, column tops were modified to include ties to beams in the longitudinal direction, creating a criss-crossing column network that enhanced stability. This modification involved not only the retention of the original architraves for horizontal connection but also the addition of ties to beams in the longitudinal direction. However, as a result of this change, *dougong* was no longer used to carry loads, became smaller, and was transformed into a purely ornamental feature in Ming and Qing architecture. Meanwhile, the hall-style structure was further simplified and ultimately transformed into a timber frame without *dougong*.

Column and tie construction. This structure employs a direct transfer of weight from the roof to the purlins, eliminating the need for lifting beams. Columns are connected by ties to form a unified structure. This arrangement offered cost savings and enhanced structural stability. The close columniation required no larger timber, but it worked well with small timber for a large house. The mesh structure also remained strong. However, this approach resulted in limited internal space, which was a trade-off for the use of so many columns and ties. This construction method became prevalent in the southern regions of China, where precipitation is more likely to be in the form of rain rather than snow. Additionally, the snow accumulation on roofs in these areas is not as significant as in northern regions.

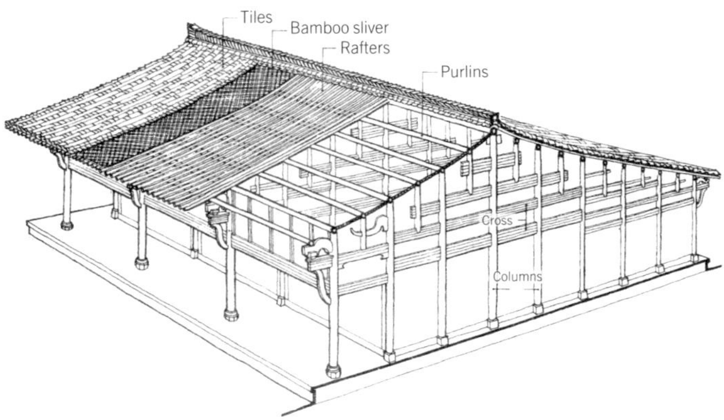

TOP Column and tie construction.

BOTTOM Log cabin construction.

In remote mountainous areas and relatively humid regions south of the Yangtze River, there was another kind of house employing a similar column-and-tie structure. First, the builders constructed a high platform on the ground floor using columns to prevent moisture and insects. They then placed beams on the platform and laid them with wooden boards, over which they built a house of column and tie style. This structure was also known as the stilt style.

In practice, due to specific conditions, mixed constructions were also employed, such as columns and ties to the gables on both sides of a house and posts and lintels on the middle beams.

Log cabin construction. The log cabin timber frame was constructed with horizontal layers of round logs or square or hexagonal timbers, cross-jointed at the corners to form the four walls of the house like a fence. The roof was then placed with dwindling timbers or short posts to connect purlins and the ridge purlin, thus creating a house.

Compared with the previous two construction methods, the log-cabin structure was less favored due to the necessity of significant quantities of wood for the construction of larger buildings, coupled with the limited range of available treatments for doors and windows. Currently, there are only a few remaining log-cabin houses in the southwest mountainous areas and the northeastern forest areas.

In addition to the three main constructions, another method for building the beam frame of pavilions was documented in *Yingzao Fashi* as the cluster angle beam structure. This method is similar to the treatment of hip rafters in the post and lintel construction, and therefore, detailed descriptions are omitted here. Another ancient structural method was the herringbone beam or diagonal beam, which is still in use in some simple and lighter houses. As the design evolved, a lower string was added to create a tripod, and extra reverse herringbone wood was placed to construct a composite tripod, which enhanced the structure's ability to withstand horizontal thrust.

Column and Post Frames

Columns were a crucial component of vertical loads in ancient Chinese buildings, making them a valuable consideration in architectural construction.

Types of columns

Columns are classified in a number of different ways. Columns are classified according to construction materials (wood, stone, or brick), shapes (straight or shuttle), cross-section shapes (round, square, or melon-shaped), and roles (interior and exterior).

Double square column, funerary ware, Wangdu County, Hebei Province

Tied bamboo column, Han tomb, Shiziwan Village, Sichuan Province

Square column, rock-cut tomb, Pengshan County, Sichuan Province

No. 1 stone column, Maiji Mountain, Tianshui City, Gansu Province

No. 30 stone column, Maiji Mountain, Tianshui City, Gansu Province

TOP Different types of columns.

BOTTOM Stone column.

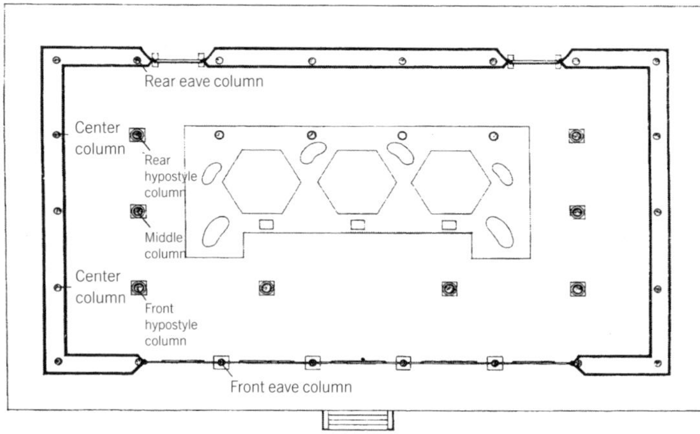

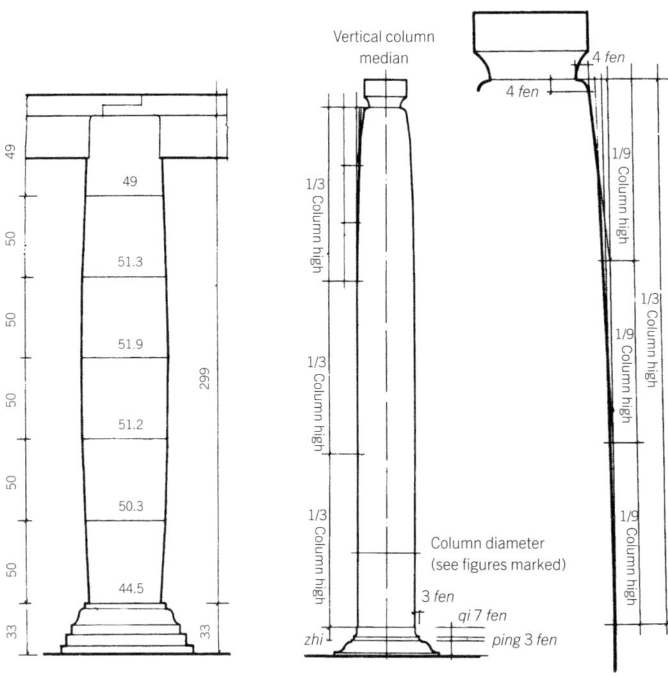

TOP Columns in different positions.

BOTTOM The way of building a melon-shaped column.

The exterior columns were exposed on the exterior of the building. The building's exterior columns were strategically placed in various locations, including under the front and rear eaves, in the center of two gables, at the four corners of the house, and above the front of doors.

The interior columns included hypostyle and center ones. The innermost pillar of a house was referred to as the inner hypostyle column, with eave columns erected between it and the outer hypostyle columns. In addition, the inner and outer hypostyle columns were divided into front and rear sections according to their positions. The middle columns were those supporting the ridge purlin. The interior columns used between the beam frame were middle posts and short posts, among others.

In addition, the towers and pavilions included a pagoda central pillar, brake cylinder, and suspended pillar.

Column treatments typically employ two methods: upper-thin and lower-thick or shuttle-shaped. The pillar's thin top and thick bottom align with the load-bearing rule and the tree's growth pattern. The shuttle pillar had two thin ends and a thick middle. The pillar appeared plump, full, and soft, reminiscent of the style popularized during the Tang Dynasty.

In multi-story buildings such as pavilions, two distinct column treatments were employed on the upper and lower floors: wrapped column construction and fork column construction. The method of wrapping the pillars involved extending the eave column on the first floor to the balcony on the second floor and adding an auxiliary pillar to the first-floor eave column to form double pillars. The additional pillar was attached to the eave column. An example of this treatment is the Ciyun Pavilion in Dingxing, Hebei Province. In contrast, the fork style involved connecting an upper column to a lower column by inserting the upper one into the *dougong* on the top of the lower column through a cross socket, with the second-story column leaning inward a certain distance.

Additionally, a unique approach was employed in the processing of columns. When the column materials were insufficient, craftsmen would use multiple timbers to assemble them into usable materials. This type of column was known as an inlay post package, which could be diversified in various ways. For instance, small pieces of wood could be wrapped around larger pieces to increase the size of the wood, or equal parts of wood could be used to join the pieces together directly. The pillars in the main hall of Baoguo Temple in Ningbo, Zhejiang Province, exemplify this technique.

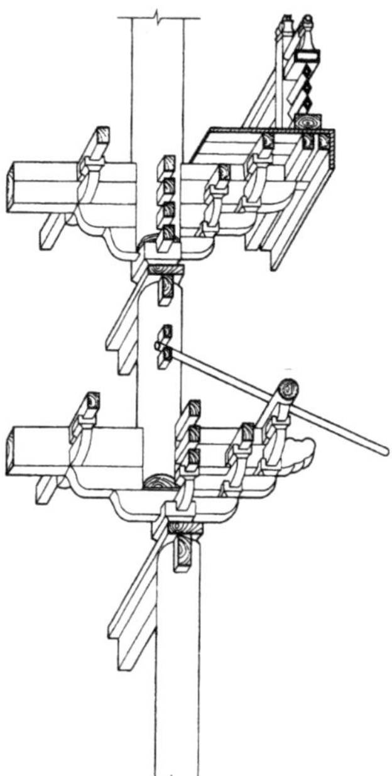

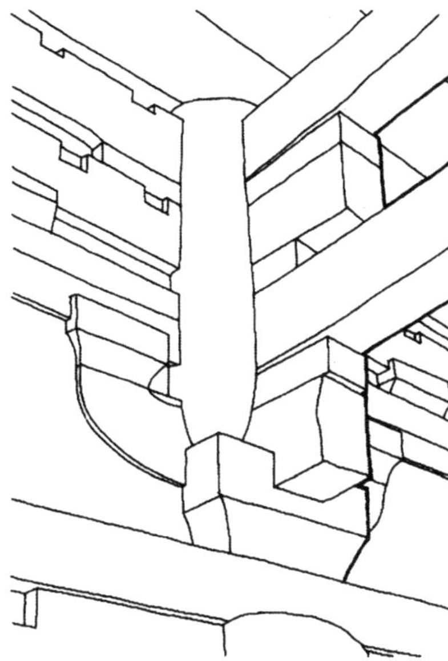

LEFT The form of fork column construction.

RIGHT The fork column construction, Guanyin Pavilion, Dule Temple, Ji County, Tianjin.

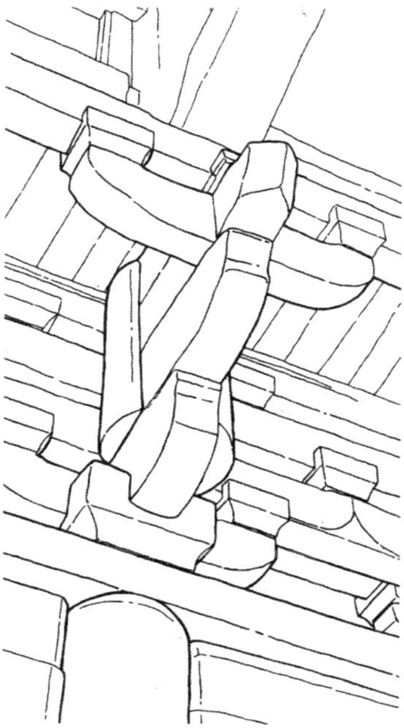

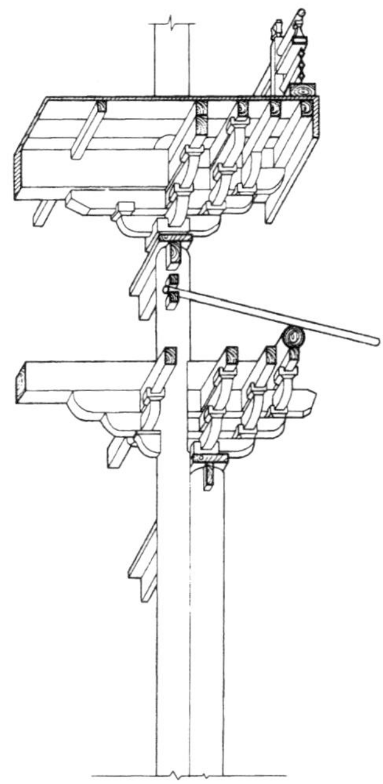

LEFT Fork column construction, Longxing Temple, Zhengding County, Hebei Province.

RIGHT The form of wrapped column construction.

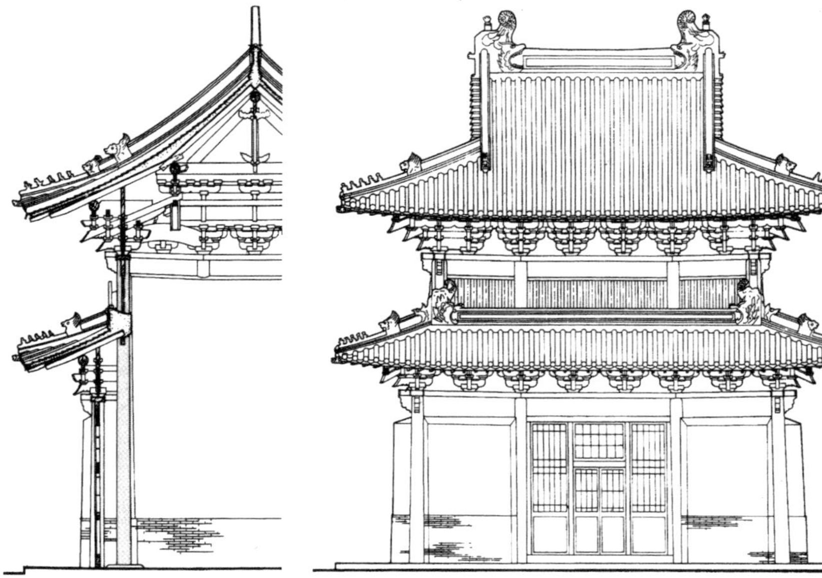

Wrapped column construction, Ciyun Pavilion, Dingxing County, Hebei Province.

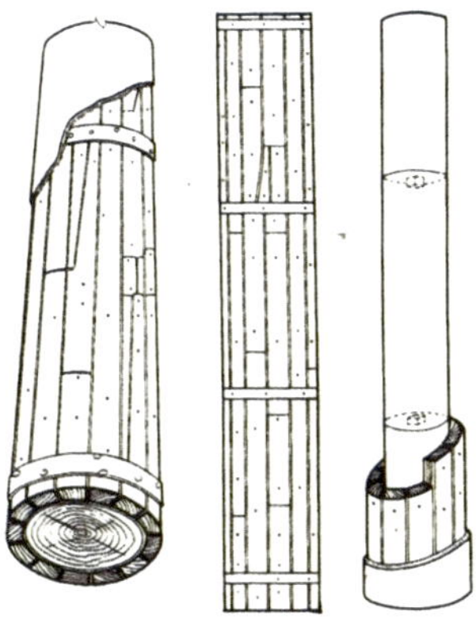

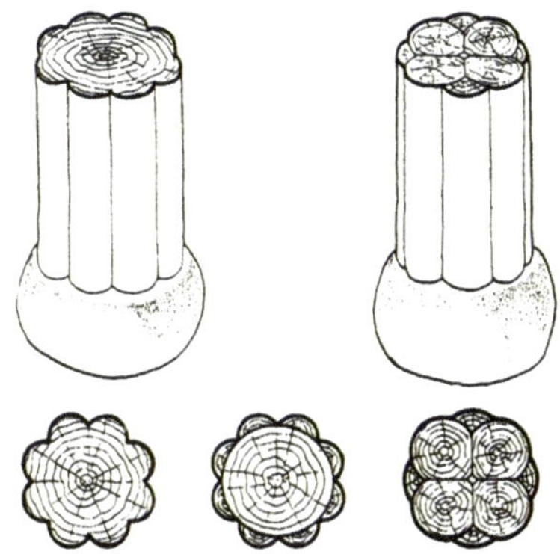

TOP LEFT Columns of inlay post package.

TOP RIGHT & BOTTOM LEFT Columns of inlay post package, Main Hall, Baoguo Temple, Ningbo, Zhejiang Province.

Placement of column foot and column base

During the excavation of early architectural sites, house columns were fixed by loading them into holes. At the Banpo site in Xi'an, there were already indications of column bases. The fire-hardened nests were wrapped around the wooden column feet. In the rammed earth platform of the Shang Dynasty, column bases were constructed using neatly arranged and well-balanced cobblestones. In the mid-Western Zhou architectural ruins in the Zhaochen Village in Qishan, Shaanxi, a large flat stone was used to support the pillars along with layered rammed pebbles as the column foundation once the basic construction of column bases was finished. The mid-Western Zhou period saw significant advances in the construction of column bases. This was due to the introduction of a new method of burial, whereby the column foot was placed from the deepest to the shallowest part of the foundation. This approach addressed the issue of wood decay caused by deep burial.

The upper timber frame underwent maturation, which strengthened its structural integrity. The base rammed earth was abandoned, and the column foot was buried from deep to shallow. In order to prevent the deterioration of wooden column feet due to moisture and decay, columns were constructed with a complete above-ground elevation and provided with support from pillar stones. Pillar foundation stones also received attention, with both their structures and forms enhanced and decorated with beautiful patterns.

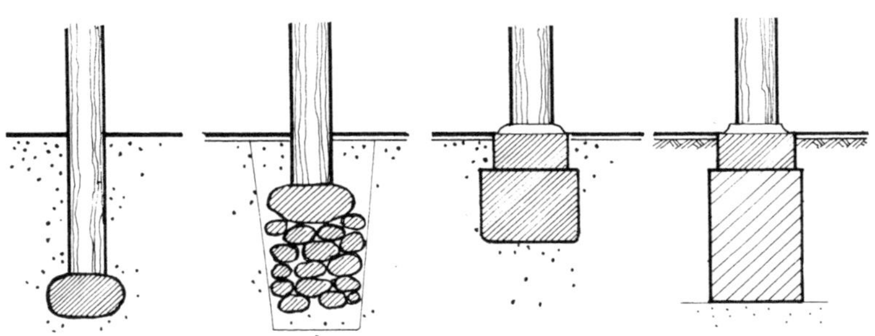

Changes of column bases underground.

***Zhi*-style column base.** *Zhi* was originally a wooden pad on the base of a stone pillar. Due to the vertical orientation of the pillar and the horizontal alignment of the wooden grains, there was a risk of moisture infiltration along the column's grain structure. To address this, a horizontal-textured wooden plate was placed beneath the column foot to act as a moisture barrier. *Zhi* was initially crafted from wood, then transitioned to metal, and eventually evolved into stone. The *Yingzao Fashi* states, "*Zhi* uses wood to replace copper." The character "櫍" (锧 or 礩) underwent changes due to changes in building materials. Subsequently, *zhi* was incorporated into the stone foundation while maintaining its original shape.

***Fupen* column base.** This plinth was transformed into a convex circle, shaped like an inverted basin called a *fupen*. There were two types of pans available: plain and patterned.

Lotus petal column base. The base was carved with intricate lotus-petal-like patterns. The upper part was encircled by lotus petals that rose upward, while the lower part was covered by lotus petals that descended downward. The majority of these designs utilized only the lower portion, also known as the ground lotus.

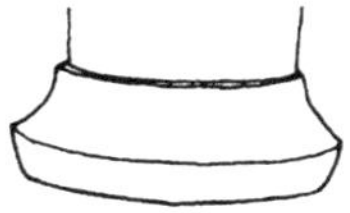

Zhi-style column base, Main Hall, Yanfu Temple, Wuyi County, Yuan Dynasty

Zhi-style column base, Main Hall, Zhenru Temple, Shanghai, Yuan Dynasty

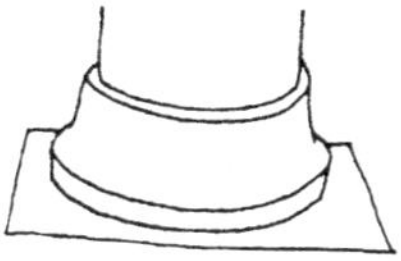

Zhi-style column base, Caomen Hall, She County, Huizhou District, Anhui Province, Ming Dynasty

TOP *Zhi*-style column base.

BOTTOM *Zhi*-style column base, Zhenru Temple, Shanghai.

Drum pier column base. One type of this base was shaped like a convex drum, called a "drum pier." Another form of stone piers was crafted into a concave drum shape, starting with a square stone that transitioned inwardly to a round shape and then connected to the column foot. Mixed moldings were processed into a circle around the periphery, circular on the plane, which was called a "drum mirror."

Fupen column base, Main Hall, Fengguo Temple, Yi County, Liaoning Province

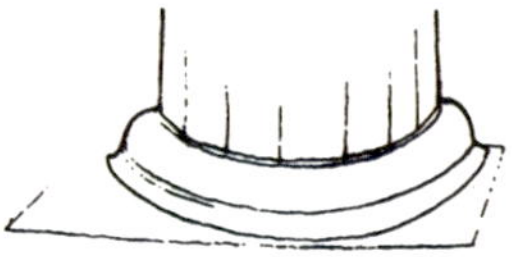

Fupen column base, No. 43 Grotto, Maiji Mountain, Tianshui City, Gansu Province

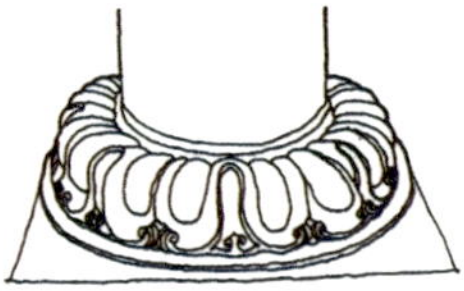

Lotus petal column base, Main Hall, Lingyan Temple, Changqing District, Shandong Province, Song Dynasty

Lotus petal column base, Dacheng Hall, Confucian Temple, Qufu City, Shandong Province, Qing Dynasty

TOP LEFT *Fupen* column base.

TOP RIGHT Square *fupen* column base, Fuxi Temple, Zhongzhang Village, Zhangzi County, Shanxi Province.

BOTTOM LEFT Lotus petal column base.

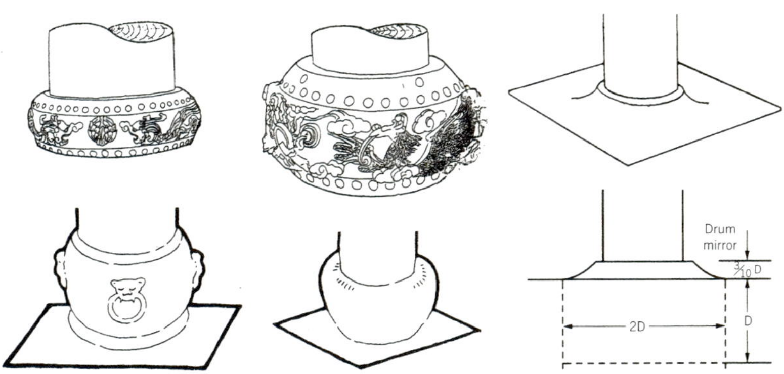

TOP Lotus petal column base, Main Hall, Foguang Temple, Wutai Mountain, Shanxi Province.

MIDDLE Lotus petal column base, Baoguo Temple, Ningbo, Zhejiang Province.

BOTTOM Drum pier column base.

Combined column base. The stone pillar base was composed of multiple types of plinths, each with its own distinctive style and flexible details.

The form and adornment of column bases developed during the process of evolution, with each dynasty demonstrating its own unique characteristics.

In the Han Dynasty, the bases were often perceived as crude and clumsy, with a resemblance to an inverted *ludou* (cap block). In the Han Dynasty, tombs and stone plinths were carved into piers or beasts.

During the Western Jin and Eastern Jin dynasties and the Southern and Northern Dynasties, the *fupen* column base was the most common type and also the basic style for later periods. With the rise of Buddhism, lotus petal bases with symbolic meanings associated with the religion began to appear. The lotus petals were narrow and long, and the overall base shape was pulled up. The prosperity of Buddhism also led to the use of animal-shaped stone foundations, including white elephants, lions, and other animals, as well as more unusual shapes, such as figures, lions, beasts, and *xumizuo*.

In the Tang Dynasty, the bases of pillars were commonly adorned with inverted lotus-shaped pedestals, which were widely used. The pedestal of the lotus flower in the Tang Dynasty was observed to be mostly circular, protruding slightly above the platform surface. Most of the lower stones were found to lie within the platform, flush with its surface.

In Song, the mainstream remained *fupen* column base, with ornamentation including lotuses, flowers, plants, dragons, phoenixes, fish, water, lions, and other patterns. The masonry system documented in the *Yingzao Fashi* included various carving techniques, including deep relief, bas-relief, flat relief, and plain line engraving. In palaces and temples, panned lotus-petal column bases continued to be a significant feature.

The plinths of the Yuan Dynasty underwent a significant transformation, moving from a cumbersome to a simple design. They were crafted without carvings or adornments and were mainly made of plain *fupen* columns. The basic styles in the Ming and Qing periods remained relatively consistent and appeared simpler than before. In more expensive buildings, drum-mirror column plinths were favored in the north, while in the south, where the climate was humid and rainy, higher drum-shaped plinths were more common. In the south, there was a greater variety of folk column bases, with intricate carvings and patterns and sometimes multiple layers.

TOP Drum pier column base, Baoguo Temple, Ningbo, Zhejiang Province.

MIDDLE LEFT Drum mirror column base, Baoguo Temple, Ningbo, Zhejiang Province.

MIDDLE RIGHT Drum mirror column base, Summer Palace, Beijing.

BOTTOM Drum mirror column base, Longxing Temple, Zhengding County, Hebei Province.

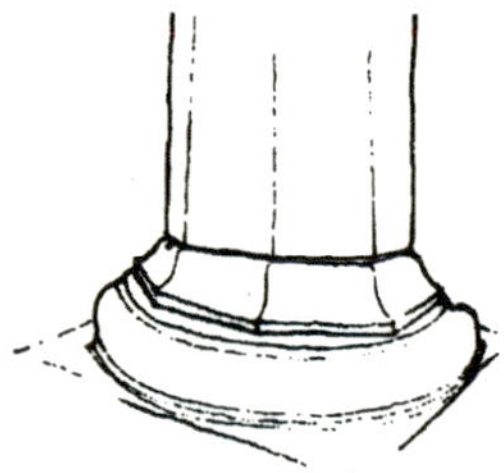

Basin-lip *fupen* column base, Xuanmiao Temple, Suzhou, Song Dynasty

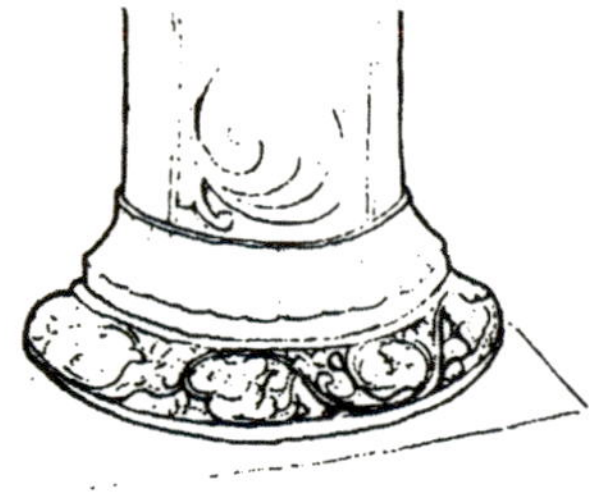

Basin-lip *fupen* column base, Arhat Temple, Suzhou, Song Dynasty

Double-layered column base of grounded lotus and curly grass, Bahui Temple, Quyang County, Jin Dynasty

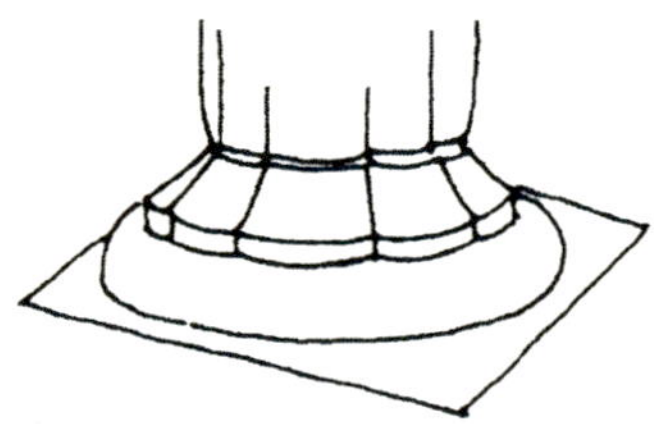

Zhi-style column base of plain *fupen* with eight petals

TOP & BOTTOM LEFT Combined column bases.

BOTTOM RIGHT Combined column bases, Sanzong Temple, Dazhonghan Village, Zhangzi County, Shanxi Province.

TOP Combined column base, Dule Temple, Ji County, Tianjin.

BOTTOM Special-shaped column base, Main Hall, Erxian Temple, Jincheng City, Shanxi Province.

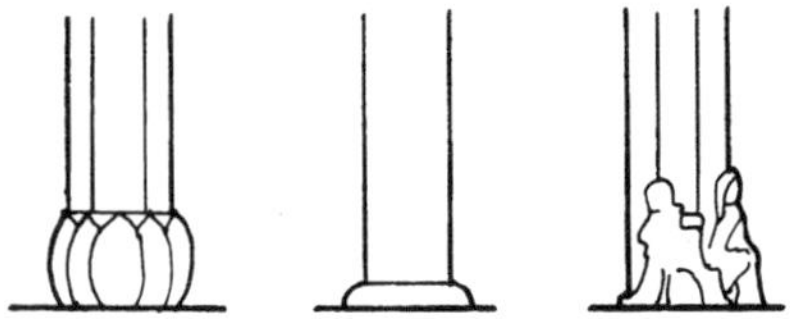

Six Dynasties
Yungang Grottoes, Datong City

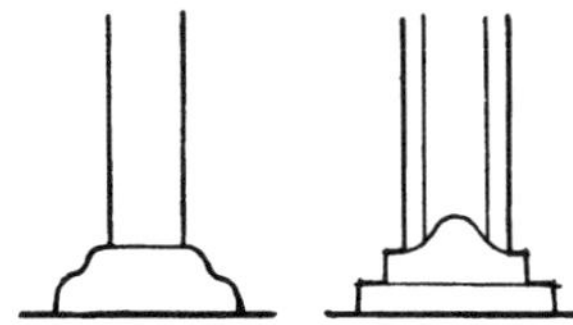

1, 2, 3, stone engravings, Wu Liang Shrine, Han Dynasty

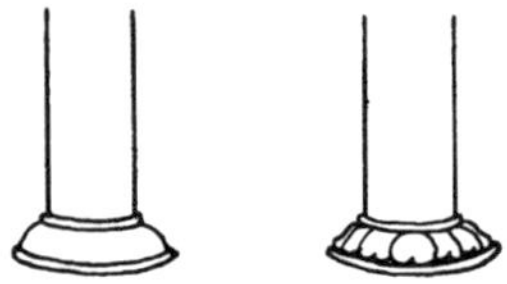

Tang Dynasty
Stone engravings on the tympanum lintel of Dayan Pagoda, Xi'an

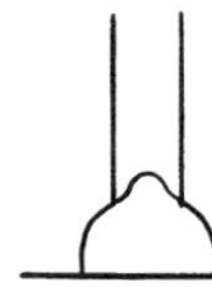

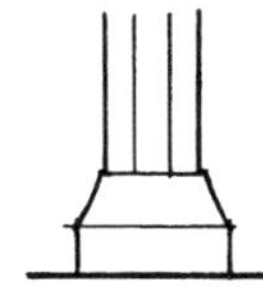

Stone Memorial Hall at Guo Ju's Cemetery on Xiaotang Mountain, Han Dynasty

To make *zhi* under a column, its circumference is 3 *fen* bigger than that of the column, and its thickness is 10 *fen*, of which 7/10 under is called *ping* and 7/10 above *qi*. The *qi* of *zhi* is reduced 3 *fen* in circumference to join and match the column body seamlessly.—Chapter 5 of *Yingzao Fashi*

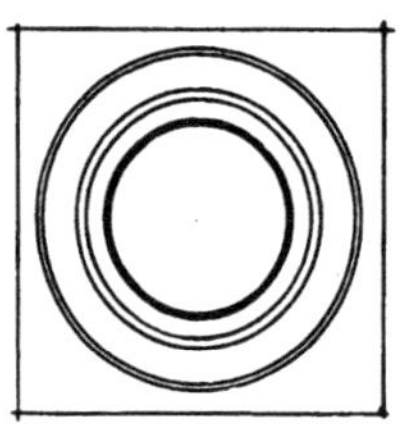

Song-style column base

The size of *zhi* is 3 *cai* of column diameter.

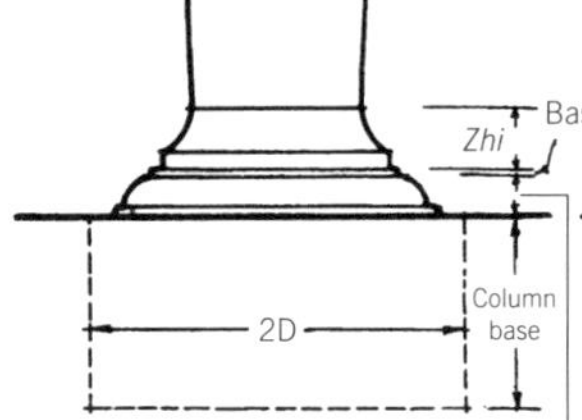

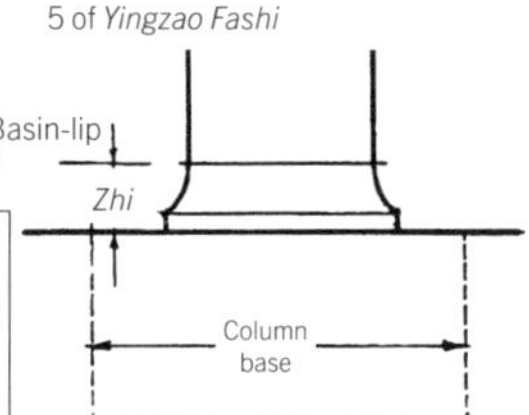

Drawn from Chapters 3 and 5 of *Yingzao Fashi*

Height of pan
1/10 2D

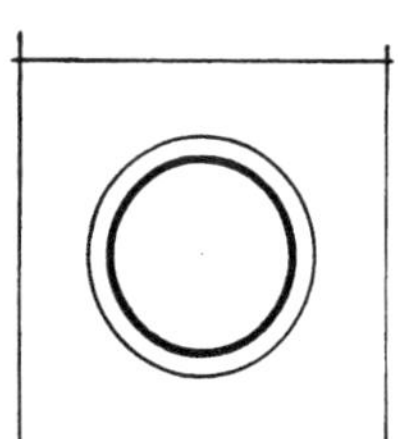

Qing-style column base

D=Column diameter

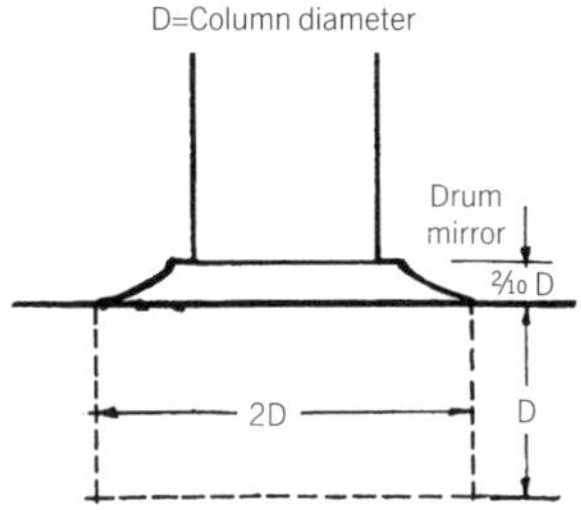

The perimeter of the column base is double the diameter of a column. For column bases with a perimeter under 1.4 *chi,* for every *chi* of the perimeter, the height of bases would be 8 *cun*; for a perimeter over 3 *chi*, the height half of the perimeter; for perimeter over 4 *chi*, the height 3 *chi*. If to make panned column bases, for every *chi* of their perimeter, the pan is 1 *cun* high and the lip of the pan 1 *fen* high. If to make upturned and downturned lotuses, its height is double that of a pan.—Chapter 3 of *Yingzao Fashi*

Yingzao Fashi: The perimeter of the column top is double the column diameter, its height is the same as the column diameter, and the height of the drum mirror is 2/10 that of the column top.

Column bases in different dynasties.

The cejiao *(sidling) of the pillars and the rise of the eaves*

In traditional Chinese architecture, columns were designed to lean slightly inward, rather than being completely vertical. Furthermore, from a comprehensive perspective, the cornice appeared to be neither horizontally straight nor perfectly vertical. Instead, it appeared to be straight in the middle but slightly elevated at both ends. These two forms were the result of the column foot being set at an angle and the eaves being raised.

Volume 5 of *Yingzao Fashi* introduced techniques for the construction of columns, "When erecting a column, it is important to ensure that the capital slightly leans inward and its foot slightly outward, which is known as *cejiao* (sidling)," "In the façade direction, the *cejiao* is 1/100 of the height of the column; in south-north or width direction, the *cejiao* is 8/1000 of the height of the column. At the corner post, *cejiao* is treated according to the above dimensions in both directions." The rise was completed in three locations: at the eaves, corner columns, and the ridge purlin. The first two rises meant a gradual increase in the height of columns from the central bay toward the corners of the building, and the rise of the ridge purlin entailed the addition of cushioning pillow wood on the purlin to make it thicker.

The techniques of *cejiao* and rises outlined in *Yingzao Fashi* were complicated and required careful consideration to ensure that a single move did not affect the whole structure. A minor alteration to the height and foot of a column would result in significant complexity in the construction of the entire timber frame, especially in the fitting of mortises into tenons. In addition to the aesthetic impact on the exterior of a house, *cejiao* also serves to satisfy the weight-bearing needs of a timber frame system.

As previously stated, the palace-style buildings of Tang and Song utilized laminated timber frames with a column layer, bracket set layer, and roof layer stacked from low to high. In accordance with the established construction methodology, the column frame was constructed first, with the columns connected to the architrave through the use of straight tenons. Once the columns were offset during construction, the column frame would collapse. To prevent this, the column heads were squeezed toward the interior to generate a push force between the architraves and the columns, thus balancing their original pull force. The greater the gravity on the roof, the stronger the pull force between

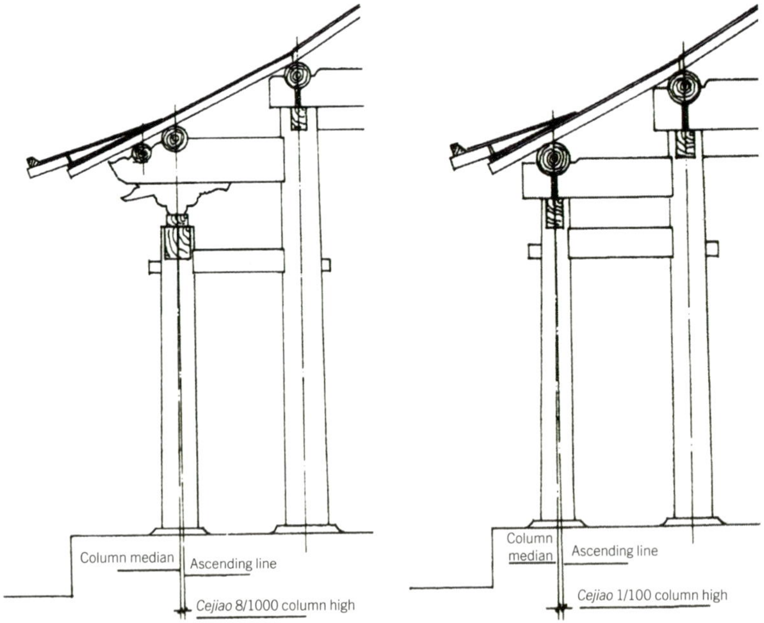

TOP Column base of Stone Shrine, Xiaotang Mountain, Shandong Province.

BOTTOM Ways to make *cejiao* column.

the columns and architraves, which makes the frame more stable and less likely to collapse. Furthermore, the rises of corner columns transformed the column capitals into a basin-like curved surface centering around the house, above the surface superposed with bracket sets and a roof. The weight of the roof increases the centripetal force of the overall timber frame, which effectively promotes the cohesion and stability of the timber frame.[4]

The remaining buildings of the Tang Dynasty exhibited the visual characteristics of upturned eaves and soft beauty. In the murals and Han portraits, the architectural image depicted straight cornices, simple and dignified. This is exemplified in the existing cornice in the Ground Stone Temple in Xiaotang Mountain, Shandong Province. A comparison of different dynasties demonstrated a maturing process of both *cejiao* and rises, as well as the timber frame systems since the Sui and Tang dynasties. In the Song, Liao, Jin, and Yuan dynasties, these two techniques prevailed, and their sizes turned out to be even larger than those specified in the *Yingzao Fashi*. During that period, wooden architecture displayed a prominent feature: the ends of architraves rose, corner columns leaned inward, and the ends of ridge purlins rose accordingly. However, as the Ming Dynasty commenced, the prevalence of *cejiao* and rises declined significantly, to the extent that Ming buildings exhibited a notable divergence from those constructed during the Tang, Song, and Yuan dynasties.[5]

Cejiao and rise in *Yingzao Fashi.*

In the early Ming Dynasty, the sideling and rises of Tang and Song were still in use and were frequently incorporated into palace buildings. However, the use of these techniques declined in the Song Dynasty. By the mid-Ming, with the exception of some folk buildings in the south of the Yangtze River, these techniques had declined in official architecture or even disappeared in some buildings. This change was closely related to the advance of the timber frame in the Ming Dynasty and beyond. After the Ming Dynasty, large wooden structures were integrated in a scientific and reasonable manner. The interior columns grew taller, and a significant number of connecting ties were used between columns. Ties to beams and penetrating ties were also widely employed. The bracket set layer was no longer required to bear loads and was therefore eliminated as an unnecessary expense. As a result, the labor- and time-consuming *cejiao* and rises were completely eliminated.

After the Qing Dynasty, the *Gongcheng Zuofa Zeli* clearly stipulated that official buildings should be designed with a *cejiao* and rise. However, these techniques were not yet in use in some timber frame pavilions and folk buildings in the regions south of the Yangtze River and were occasionally observed in official architecture. This was probably due to construction habits, but they were only applied to exterior eave columns, not to interior columns.

Column network, column reduction, and column shifting

A Chinese wooden building with multiple bays must establish a column network, that is, the plan layout of all columns in a house.

This column network in Chinese timber frame differs from that in modern architecture. In contemporary construction, columns are erected at the vertical and horizontal intersections of a house structure to form a grid. In traditional timber-frame buildings, a network is constructed by arranging columns at the crisscrosses of bays so that sufficient space is left below the beams. If the modern approach were employed, Chinese wooden houses would be filled with columns, leaving narrow usable space.

Yingzao Fashi described the four layout methods of the column network for palaces, also called the four *cao* styles. *Cao* could refer to a deeply dug base, equivalent to the foundation excavation for modern buildings, on which columns were arranged to make all column bases identical. Another explanation of *cao* points to the location of columns on the plane. The four types of *cao*s excluded the side corridors, colonnade and rooms around the principal bays. Except for the *fenxin doudi cao*, a palace used eight-rafter-beam with seven bays in length and four bays in depth, the inner space was two bays of four-rafter-beam in width, and the main beam employed four-rafter-beam plus a two-step cross beam. This proved to be the most economical and reasonable structural form.

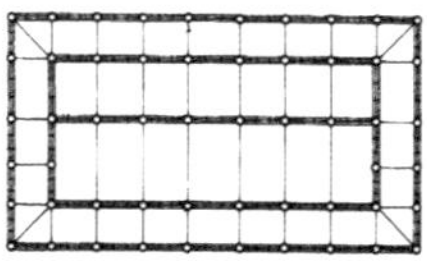

Single *cao*—the foundation of halls in *Yingzao Fashi*

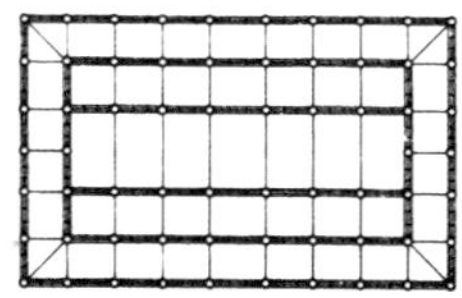

Double *cao*—the foundation of halls in *Yingzao Fashi*

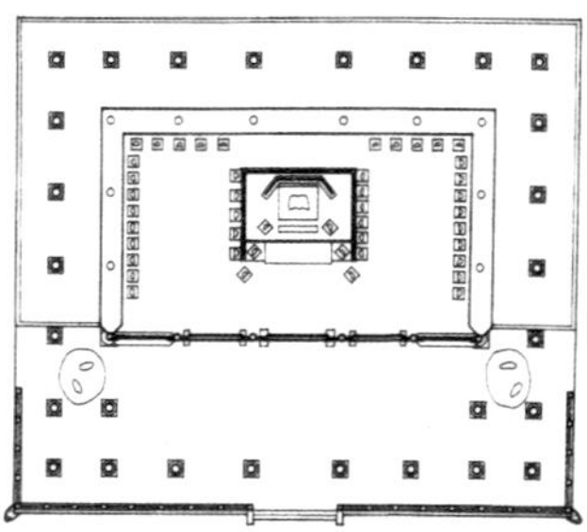

Plan of the Shengmu Hall, Jinci Temple

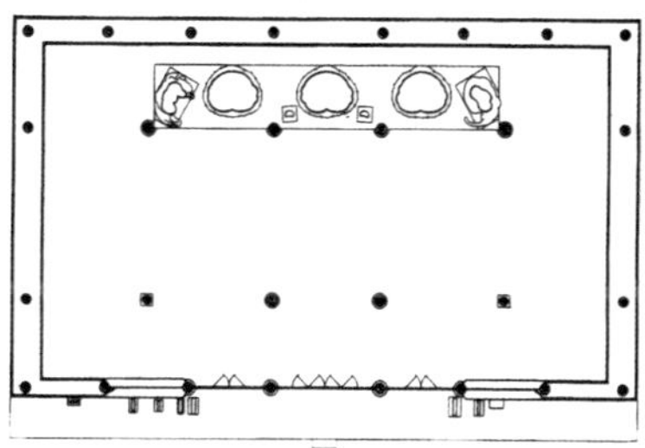

Plan of Rear Hall, Guangsheng Lower Temple, Hongdong County, Shanxi Province

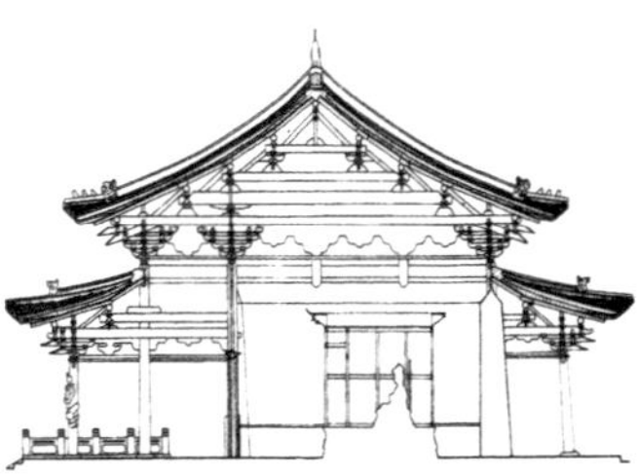

Cross section of Shengmu Hall, Jinci Temple

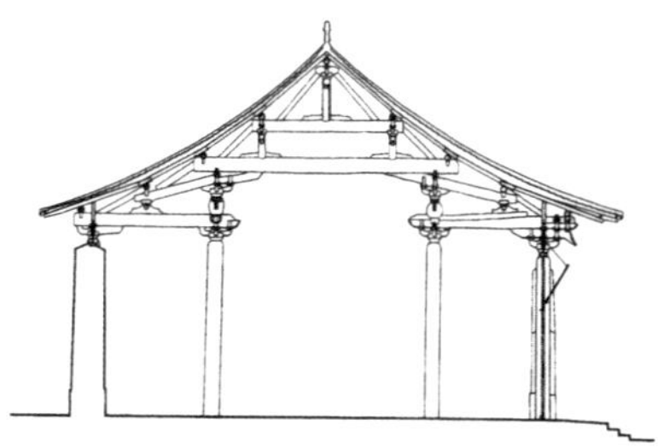

Cross section of Rear Hall, Guangsheng Lower Temple, Hongdong County, Shanxi Province

LEFT Single *cao* (Shengmu Hall, Jinci Temple).

RIGHT Double *cao* (Main Hall, Guangsheng Lower Temple, Hongdong County, Shanxi Province).

Single *cao*. The palace was seven-bayed with two rafters for side corridors and rooms around bays, its plan form being a rectangle around exterior eave columns. The interior bays were divided by a row of columns, with the indoor space divided into a single *cao* about two rafters deep and an outer *cao* about four rafters deep.

Double *cao*. The palace was seven-bayed with two rafters for side corridors and rooms around bays, its plan form being a rectangle around exterior eave columns. The interior bays were divided into two sections by two rows of columns. The front section was two rafters deep, while the rear section was four rafters deep. The middle section was four rafters deep.

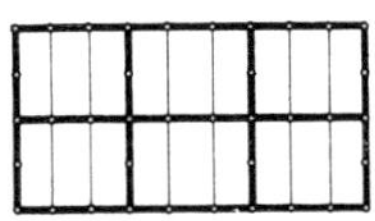

Fenxin doudi cao—the foundation of halls in *Yingzao Fashi*

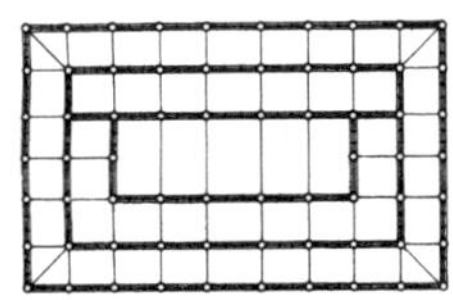

Jinxiang doudi cao—the foundation of halls in *Yingzao Fashi*

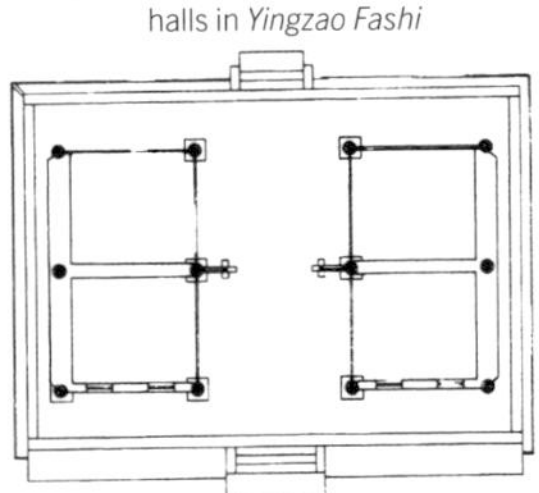

Main Gate, Dule Temple, Ji County

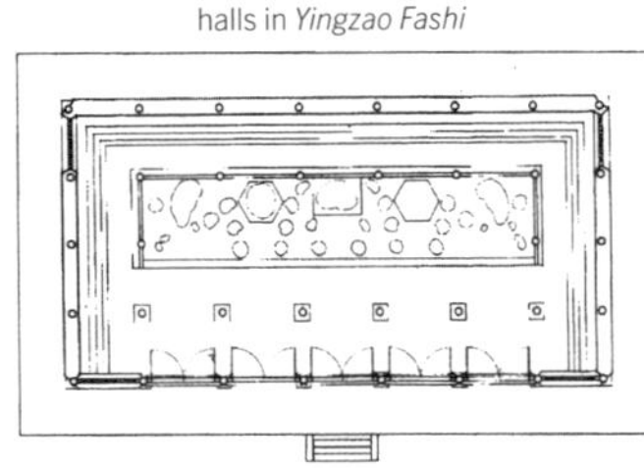

Plan of Main Hall, Foguang Temple, Shanxi Province

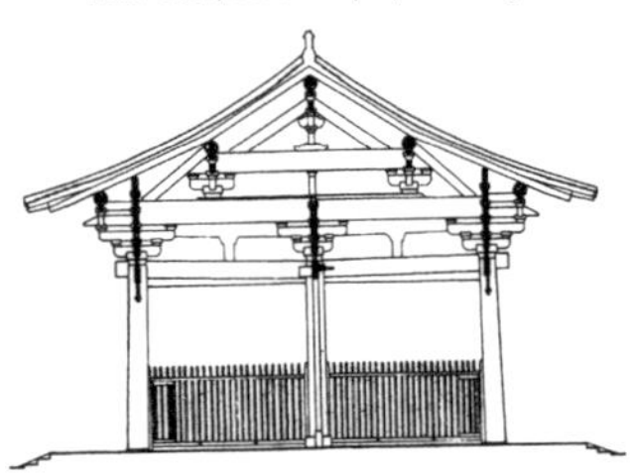

Cross section of Main Gate, Dule Temple

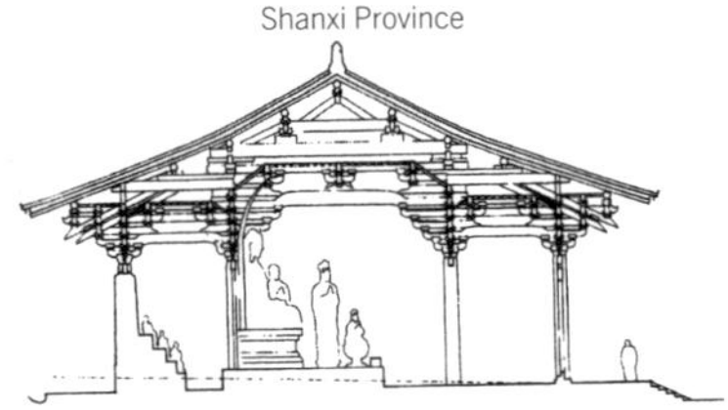

Cross section of Main Hall, Foguang Temple, Shanxi Province

LEFT *Fenxin cao* (Main Gate, Dule Temple, Ji County, Tianjin).

RIGHT *Jinxiang doudi cao* (Main Hall, Foguang Temple, Shanxi Province).

Fenxin doudi cao. The palace was nine-bayed. The plan was a rectangle surrounded by eave columns, with a vertical row of columns dividing the indoor space into two equal parts within each bay. The palace had nine bays, with a horizontal row of columns separating every three bays laterally to form a *fenxin doudi cao*. Among the four types of *cao*, only this type has nine bays without side corridors and rooms around bays, while the others are all seven-bayed. Among extant timber buildings, the nine-bay *fenxin doudi cao* is a rare occurrence, while the three-bay and five-bay are more commonly seen.

Jinxiang doudi cao. The palace was seven-bayed with two rafters for side corridors and rooms around bays. The plan form was two rectangles, one formed by the eave columns and the other by the interior columns, with two rafters positioned away from each other. In the rear row of columns, architraves and plates were connected to the eave columns. From the perspective of mechanics, *Jinxiang doudi cao*, which resembles the Chinese character "回," is the most stable of all the four *cao*s.

The four types of column network were arranged around the middle of a palace to create a complete plane space, its length varied with the bays, while its width or depth was always two bays and four rafters. This was the main plane space constituted by the beam frame, which serves as the venue for living activities. This is what Mr. Zhang Jiaji considered the concept of *doudi*.[6]

Occasionally, the architectural space derived from a column network required adjustment. In such instances, the layout of the columnation had to be modified by reducing or relocating columns. A reduction in the number of columns was achieved by removing some, while a relocation attempted to move the remaining columns to create the desired new space. Both reduction and relocation adjust from the original columniation or the prescribed column network.

From the architectural relics, reduction, and relocation were concentrated in Liao, Jin, and Yuan. Some people believe that these adjustments came into use in the Central Plains, where ethnic minorities dominated and settled for three generations. When absorbing the architectural culture of the Han nationality, these minorities tailored their house construction to their traditional habits. Column reduction and relocation could be employed independently or in conjunction with one another. For instance, the Sansheng Hall of Shanhua Temple in Datong, Shanxi Province, features four bays of length, with the removal of all interior columns in the front eaves and the side bay on the back eave, which has moved the inner columns of one rafter inwards.

In the Ming and Qing dynasties, these two techniques were no longer employed due to their complexity and the inherent risks associated with irregular beams and long-span space structures. It was only a matter of time before this practice declined and disappeared.

Fang (Tiebeam) and Sparrow Brace

A tiebeam is a connecting component in a timber frame that enhances its stability. The name of this product varies depending on the location, dynasty (Song to Qing), and region.

E'fang *and* pupaifang

E'fang was a horizontal linking tie between the upper ends of columns. It was named *lan'e* during the Song Dynasty and called *e'fang* during the Qing Dynasty. *Yingzao Fashi* set forth the specifications for its production, "To make *lan'e*, the width of the material was doubled, but the thickness was reduced by one-third, and the length varied with the spacing of the width, and was halved from the ends to the center of the columns, and thence to the dowels of the posts. Each shoulder is ornamented with four petals, each of which is one-eighth the length. If no padding is used, half the thickness of the width is taken out." The high cross-section of the appendix for the doubling of the material is 30, with a thickness of minus 1/3 of the high, that is, 20. The length is in accordance with the width of the two ends, which are inserted into the center of the column. The mortise and tenon column should be half the thickness of the square, or 10. The two shoulders of the rolling head should be cut into the curved shoulders, as the appendix on the forehead is not applied to the arch. The cross-section thickness should be 15.

TOP *Lan'e* in Nanchan Temple, Wutai Mountain, Shanxi Province.

BOTTOM *Lan'e* in East Hall, Foguang Temple, Wutai Mountain, Shanxi Province.

Prior to the Southern and Northern Dynasties, *Lan'e* was typically positioned at the top of columns. However, following the Sui and Tang dynasties, they were relocated to the tops of beams and columns. In larger buildings, it usually contains two layers. In the Song Dynasty, the upper layer was called *lan'e*, and the bottom layer was known as *you'e* (sub-lintel). In the Qing Dynasty, the upper layer was designated as the "big *e'fang*," while the bottom layer was referred to as the "small *e'fang*." The sub-lintel served a dual purpose, both as an embellishment and as a structural component. A thin board called *you'e* cushion board in the Song Dynasty was placed between the lintel and the sub-lintel. According to *Yingzao Fashi*, "Sub-lintel is set below the lintel, and has a cross section that is 1/2 or 1/3 shorter than the lintel, with a length of 15 or 20 *fen*.* The regulations for mortise and entasis of the sub-lintel are identical with those of the lintel."

From the earliest buildings to the Tang architecture, the cross-sectional height-to-width ratio of the *lan'e* was approximately 2:1, with no protrusions for corner columns. The Liao *lan'e* exhibited a protruding design at the corner columns and was cut vertically, in keeping with the Tang style. The ratio was changed to approximately 3:2 until the Song and Jin dynasties. The Song style *lan'e* featured protrusions in the form of *shuatou* (head of the beam), while the Jin style *lan'e* exhibited similar protrusions or *bawangquan* (*Bawangquan* has three convex semicircular lines in the middle and two concave semicircular lines at both ends, which are continuously formed into a flower-shaped head decoration). The Yuan style *lan'e* resembled a petal pattern.[7] The cross section of the Ming and Qing buildings became similar to a square, with a ratio between height and width that reached nearly 1:1. Protrusions were largely in the *bawangquan* style. In some folk buildings, the protrusion was not readily apparent, and even in the Ming and Qing periods, vertical sections of Liao were still in use. In fact, this protrusion was beneficial for the stability of a timber frame.

E'fang connected columns so that a complete column frame was built stably while at the same time sustaining *dougong*s between columns. Early buildings had bracket sets on columns, and later additional sets were installed between

---

* Please refer to the *cai-and-fen* system in the "*Dougong*" section below.

Evolution of *lan'e* and *pupaifang.*

columns to demand more weight bearing ability from *e'fang*, so *e'fang* existed not only to decorate but also to carry loads.

Pupaifang resting on *e'fang* was another form of small beam, and its original application was to undertake *dougong*. Besides being more conducive to placing and bearing *dougong*, the plate had an even greater impact on the later use of diagonal bracket-arm. Larger in shape and with more weight to be carried, diagonal bracket arm would deform with only a lintel but no plate.

Pupaifang was placed on an *e'fang* and received the tiebeam of the cap block (the principal bearing block, lowest in a bracket set) of *dougong*, called the wood-plate lintel in the Qing Dynasty. According to *Yingzao Fashi*, Song palaces employed lintel alone to support the *dougong*, but the pavilion bracket set supplemented a plate that was 15 to 21 *fen* thick and used square wood in the width of the cross section. If wrapped column construction was applied, below the plate, there would be corner columns 45 *fen* wide and 30 *fen* thick with mortises on the top.

Pupaifang was first seen in the Tang Xuanzang Brick Pagoda, imitating a wooden frame, in Xingjiao Temple in Xi'an, but it was not used in Nanchan Temple or Foguang Temple of Wutai Mountain, which are the wooden architectural remains from the Tang Dynasty. *Pupaifang* was more popular during the Liao and Song dynasties, although it was absent in some buildings. After the Jin Dynasty, *pupaifang* became a standard component of architectural design. The cross-section of *pupaifang* was initially wide and thin, gradually thickening over time. By the Ming and Qing dynasties, the cross-section was narrower than *e'fang*. In the early stages, the protruding element did not extend to the corner columns. However, this changed later, with the protruding element being cut vertically. Begonia-style protruding heads were introduced in the Yuan period.

In addition to the *e'fang* and *pupaifang*, a horizontal bracket-arm featured a horizontal lintel centered on the column capital. A *zhutoufang* was located on the axial bracket arm. A *luohanfang* was positioned on an oval or long arm. Finally, a *liaoyanfang* was situated on a regular arm under the exterior eave.

TOP *Lan'e* and *pupaifang*, Kaishan Temple, Gaobeidian City, Hebei Province.

FOLLOWING PAGE *Lan'e* and *pupaifang*, Yanqing Temple, Wutai Mountain, Shanxi Province.

Pupaifang
Lan'e

Pupaifang
Lan'e

PREVIOUS PAGE *Lan'e* and *pupaifang*, Wenshu Hall, Foguang Temple, Wutai Mountain, Shanxi Province, Jin Dynasty.

TOP *Lan'e* and *pupaifang*, Scripture Library, Qinglian Temple, Zezhou County, Jincheng City, Shanxi Province.

Lan'e and *pupaifang*, Chunhua Temple, Pingshun County, Shanxi Province, Jin Dynasty.

TOP *Lan'e* and *pupaifang*, Yuanjue Hall, Faxing Temple, Zhangzi County, Shanxi Province.

BOTTOM *Lan'e*, Dule Temple, Ji County, Tianjin.

TOP *Pingqifang*, Foguang Temple, Wutai Mountain, Shanxi Province.

BOTTOM *Lan'e*, *pupaifang*, *zhutoufang*, *luohanfang*, *liaoyanfang*, Fotou Temple, Pingshun County, Shanxi Province.

Lan'e, *pupaifang*, *zhutoufang*, *luohanfang*, *liaoyanfang*, Zhengjue Temple, Zhangzi County, Shanxi Province.

Luohanfang
Zhutoufang

Sparrow brace

The sparrow brace is installed between columns and tiebeams at the corner of the structure to enhance the load capacity of the tiebeams, strengthen the connection, and shorten the net span of the tiebeams. As it could prevent the intersection angle from tilting, it was also known as a corner brace. The Sparrow Brace, which resembles a bird's wing on a column, is also known as the *chuomufang*. This term was used during the Song Dynasty.

It is speculated that the sparrow brace originated from the *timu*,* and its image was similar to the *shipai gong* (located in the cap block to receive the flower arm and axial bracket arm). Its earliest image appeared in the relief sculpture in Cave 8 of Yungang Grottoes, built in the Northern Wei Dynasty. In the main hall of Kaishan Temple in Xincheng, Hebei Province of the Liao Dynasty, a sparrow brace was discovered. It was composed of two layers of *shipai gong* (bracket arm), which is consistent with the content of the stone carvings of Yungang Grottoes.

No evidence of a sparrow brace was discovered in Tang, and specimens without the use of a sparrow brace in some high-level buildings existed in Song, Liao, Jin, and Yuan.[8] In the Song *Yingzao Fashi*, the front and lower parts of the sparrow brace were of two types: *tatou* (only two or three lines are carved at its ends to form several petal-like patterns) and the cicada belly. The cicada belly was the most prevalent style in Liao and Jin, with Jin's style being the most intricate, and in Ming and Qing, the style turned simpler. The form of sparrow brace was established in the Ming Dynasty. It began with *tatou*, then progressed to *xiaohun*, and finally ended with the cicada belly, with *gong* (bracket arm) installed beneath. In Qing, the dimensions of *tatou* and the cymas were increased while the cicada belly was reduced in size. In addition, starting from the Ming, the olecranon-shaped protrusion emerged on the front of the sparrow brace and remained prevalent in the Qing Dynasty. A multitude of carvings and paintings were employed during the two dynasties.

* Placed in flower arm to receive eave purlin or a narrow member inserted in the *ludou* as an auxiliary half *gong* (bracket arm) to support the bracket arm; also sometimes functions as an arm above a bracket set.

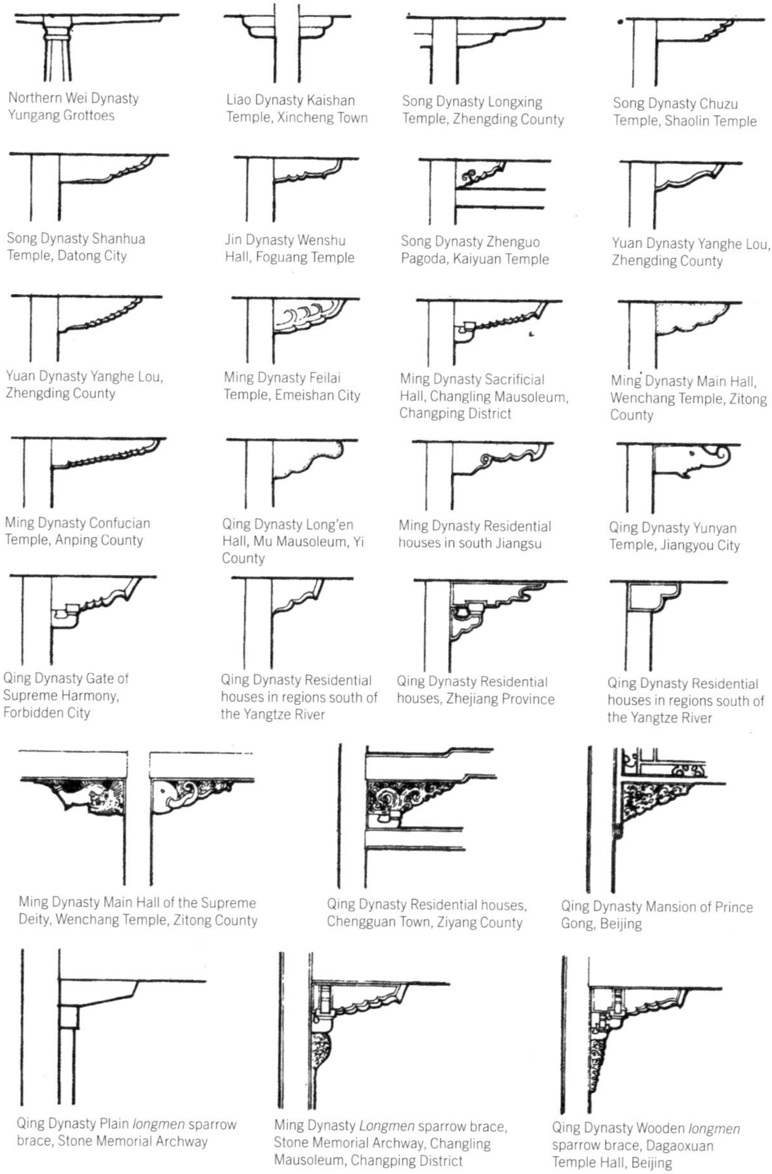

Evolution of sparrow brace and *huayazi*.

In its early days, the sparrow brace was designed to bear weight, and its shape looked plain and simple. Later, the structural function of the sparrow brace declined, while its decorative value and patterning increased.

Big sparrow brace. The term "big" in this context does not refer to a large size, but rather to the practice itself. This type of brace is constructed from a single piece of wood. It features a wide upper section that narrows downwards, with a large block beneath resting on the column capital. Columns and sparrow braces were not connected to each other, nor were column heads and sparrow braces (similar to *dougong*). Instead, they were joined together to carry the load of the roof. The big sparrow brace first appeared in the Northern Wei Dynasty and was subsequently discontinued, with the exception of its use in Tibetan Buddhist buildings.

***Tong* sparrow brace.** Two sparrow braces were installed on either side of the column shaft to form a unified structure. The upper end of the brace was flush with the column capital. In comparison to the larger sparrow brace, the tong style was positioned at a lower level. This brace was typically used outdoors and was prevalent in the Song and Yuan periods, but became less common in the Ming and Qing dynasties.

Sparrow brace. This commonly used brace was also known as the single-winged sparrow brace, while the smaller one was referred to as the small sparrow brace. From an aesthetic perspective, the two single-winged braces at either end of a column appeared to be tong-style, but in fact, they were independent of each other.

Horse-riding sparrow brace. The majority of instances involved the use of this architectural element in the terminal room or colonnade of a building. Due to the close proximity of the two columns, two sparrow braces were used to span across them, serving more as an ornamental structure than a functional one.

***Longmen* sparrow brace.** This sparrow brace was used exclusively on free-standing high gateway joined overhead. It added ornamental components such as *yundun, sanfuyun, zikuang,* and *mayetou. Sanfuyun* (cloud-cluster decoration) was installed on both sides of the brace in the width direction of a building, resulting in spatial changes. *Yundun,* put under the brace, was to support the brace while the long *zikuang* joined the column to hold the *yundun* in place.

Another sparrow brace, purely decorative and intended for hanging, was called *huayazi*. Less than four centimeters thick, this brace was carved or painted with intricate and diverse patterns, including curly grass, peonies, gourds, plum bamboo, sunflowers, and symbols for longevity.

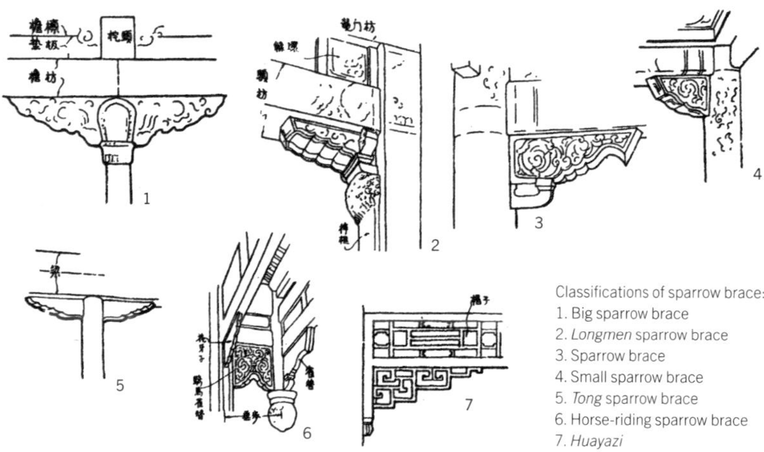

Classifications of sparrow brace:
1. Big sparrow brace
2. *Longmen* sparrow brace
3. Sparrow brace
4. Small sparrow brace
5. *Tong* sparrow brace
6. Horse-riding sparrow brace
7. *Huayazi*

Different types of sparrow braces.

Qing sparrow brace.

FOLLOWING PAGE Qing horse-riding sparrow brace.

Qing *huayazi.*

Dougong

Dougong, as one of the external visual features essential and unique to Chinese architecture, is an exclusive member of timber frame buildings. It also becomes a scale standard in Chinese architecture and a symbol of a building's grade, specification, and status. Artistically or technically speaking, it suffices to symbolize and represent the soul of Chinese classical architecture.[9] Along with the development of Chinese timber frames, *dougong* performed differently across dynasties.

The name of *dougong* was in wide use until the Republic of China. Its appellations varied in the two famous books regarding Building Construction Codes in ancient China. *Yingzao Fashi* in Song named it *puzuo*, and *Gongcheng Zuofa Zeli* of the Qing Dynasty called it *douke*.

According to its locations, *dougong* was named differently. Generally speaking, in terms of the scope of use, it was divided into an inner or an outer eave bracket set while the location of use classified it into a bracket set on columns, a bracket set between columns, or a bracket set on corner.

The evolution of dougong

The image of a *dougong* was first seen on a bronze vessel named *ling-gui* in the early Western Zhou Dynasty. At the feet of the vessel sits a cap-block-like structure to undertake the upper weight; in the cap block between the feet is a component similar to a tiebeam for connection; on the tiebeam are two squares to receive the upper part like a small block. Some experts believe that it represents the image of *dou* (bracket) in *dougong*.[10]

Dougong in the Zhou Dynasty was mentioned in some documents. *Erya · Shigong* recorded, "*Mangliu* refers to the girdle, on which *ying* means *zhuo* (king post), on which *bian* refers to *ji* (square wood), and *er* is *jie* (*dougong*)." *Er* and *jie* are the aliases of *dou* (bracket). As the earliest dictionary in China, *Erya* was written during the Warring States Period or between the two Han Dynasties. The upper limit could not be earlier than the Warring States Period because the materials in the book come from *Chu Ci* (*Songs of Chu*) and *Lüshi*

FOLLOWING PAGE Bracket sets on columns, between columns, and on corners, East Hall, Foguang Temple, Shanxi Province.

TOP Bracket sets on columns, between columns, and on corners, Fotou Temple, Pingshun County, Shanxi Province.

BOTTOM Bracket sets on columns, between columns, and on corners, Front Hall, Cuifujun Temple, Zhangzi County, Shanxi Province.

Chunqiu (*Master Lü's Spring and Autumn Annals*), both of which were written after the Warring States Period; the lower limit is not later than the Western Han Dynasty, for Emperor Wen of the Han Dynasty already assigned scholars to study *Erya* and the time of Emperor Wu witnessed the publication of *Erya Notes*. The entries and explanations of "*er* is *jie*" in *Erya* also proved that *dou* (bracket) before Han was called *er* and *jie* in Han. *Yi Zhou Book · Volume 5 · Zuoluojie*, composed in the pre-Qin Dynasty or in the Western Han Dynasty, recorded that Emperor Zhou built a second capital in Luoyi, which "constructs five palaces: an ancestral temple, an ancestral palace (royal family temple), a mourning hall set up by the emperor in the palace, the main hall of emperors and princes, and a palace for an emperor to declare punishment and prize or instructions. All buildings are characterized by the eaves on the four sides, allowing water to flow down from all sides, earthen goblet stands where toast glasses were placed to show mutual respect, overlapping roof beams, double-eave roofs, overlapped roofs and beams, double bracket sets, pillars with alga-figure, small houses attached to a pavilion or next to the main hall, multiple columns at the front of the hall, colored drawings on the caisson ceiling, and a painted door screen." Among them, double bracket sets and pillars with an alga-figure are suspected to be the image of *dougong*. According to these documents, *dougong* prevailed in Zhou architecture.

In *The Analects of Confucius · Gong Zhi Chang*, Confucius said, "Zang Wenzhong, a senior official in the state of Lu, built a divination house for the big tortoise (for divination) from Cai. The house features a bracket set carved in the shape of a mountain and beams and columns painted with algae. How can such a person be considered wise?" Confucius criticized Zang Wenzhong for violating etiquette by using decorative *dougong*. Among the cultural relics unearthed from the Warring States Period, a square copper case with a dragon and phoenix seat from the tomb of King Zhongshan in Pingshan County, Hebei Province, had four corners strutted by a *dougong* made of one cap block and two small blocks, exhibiting clear images of cap block, small block, *gong* (bracket arm), and short columns. The same clarity was demonstrated in *dougong*, depicted by the pattern of a Warring States lacquer plate excavated from Linzi, Shandong. In Warring States buildings, *dougong* was used in a comprehensive manner, indicating its extensive application, primarily for cornice support. However, there was no clear proof of its jumps, either from documentary records or physical remains.

Dougong in Qin was still single-layered but developed greatly in the Han Dynasty, as evidenced by numerous Han stone images, murals, burial vessels, and palaces. The forms of *dougong* varied, including one cap block plus two small blocks, one cap block plus three small blocks, or one cap block plus four small blocks. They were single-layered or multi-layered, and the head of *gong* was available in diverse shapes. Inverted-V *dougong* was widely used. Besides, the *gong* head underwent significant changes, with herringbone *gong*s becoming the dominant style.

The extant Gaoyi Stone *Que* in Ya'an, Sichuan, and the Wuming *Que* in Qu County had *dougong* imitating wood structures. The pottery tower unearthed from the Dawanzui Han Tomb in Leshan and the Mumashan Han Tomb in Shuangliu County, Sichuan Province, exhibited a more complex *dougong*. The stone structure of the tomb chamber in the Han tomb in Yinan, Shandong, also clearly displayed the shape imitating the wooden *dougong*. The prevalence of *dougong* in architecture gained support from the portrait stone images excavated from Shandong, Sichuan, and Xuzhou. Based on the available evidence, it can be concluded that the Han *dougong* exhibited all the essential characteristics of the full-fledged form, although it was not yet fully mature."[11]

In 2013, the Jiuyuan Gang mural tomb of the Northern Dynasty was unearthed in the northeast of Xiashe Village, Lancun Town, Xinzhou City. The tomb passage was decorated with a mural above the door. The main hall was three bays wide and roofed with a single eave and five ridges. The *dougong* was constructed in great detail, featuring two jumped *dougong*s and two sets of diagonal *dougong*s. The column capital rests on a cap block characteristic of *min* (a square board beneath the cap block) *dou*, and *dougong*s on all the eave columns stand side by side with the central bay. The cap block is placed directly on the column capitals, and the cap block on the left side has distinctive characteristics of *min dou*. As seen from the Jiuyuangang murals, in the Northern Dynasties, *dougong* already possessed complex structures, and the diagonal *dougong*s were used in a variety of ways.[12]

By the time of the Sui and Tang dynasties, the fabrication of *dougong* was well-established. Among the several existing timber frames constructed during

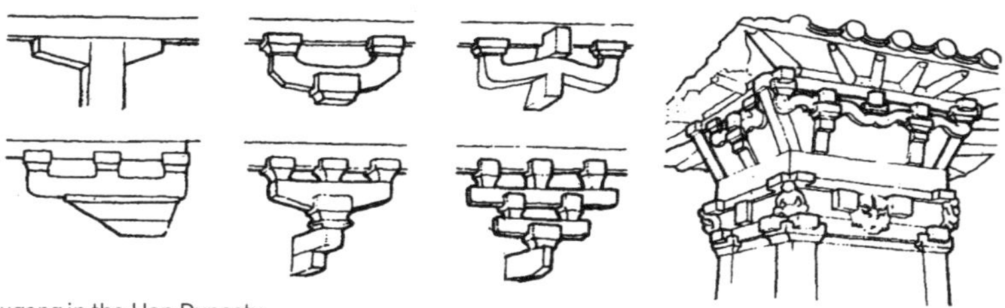

Dougong in the Han Dynasty.

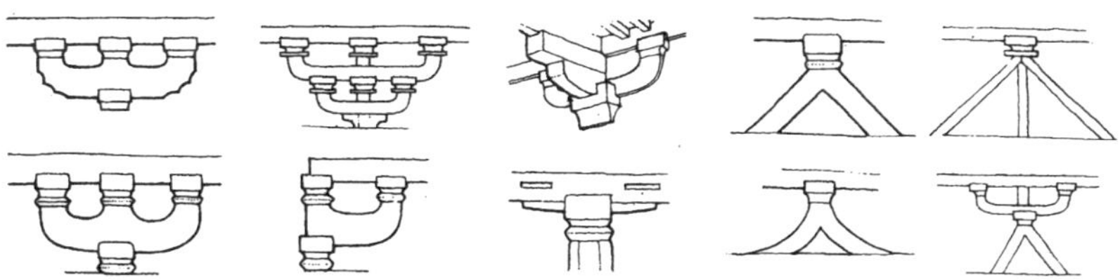

Dougong in the Southern and Northern Dynasties.

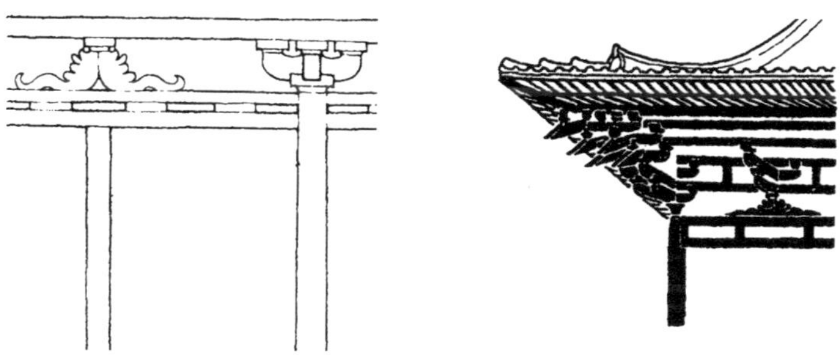

Dougong in Sui and Tang dynasties.

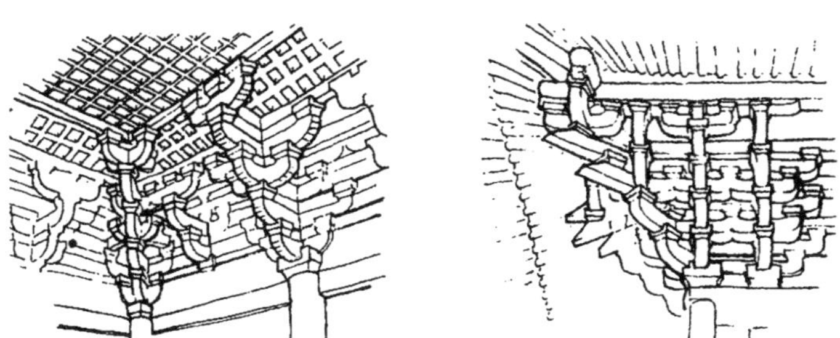

Dougong in Song, Liao, and Jin dynasties.

the Tang Dynasty, *dougong* assumed the role of supporting the weight of the roof. From relics, bracket sets between columns looked slightly simple, akin to those of the Han and Southern and Northern Dynasties. The majority of these structures used an inverted-V *gong*, or one single cap block and three small blocks. Buildings appeared grand and magnificent with the skilled production of bracket sets on columns, between columns, on the corner, and indoors.

The Song *dougong* grew more standardized. Its members, including blocks, bracket-arms, *ang*s (a long-slanted lever arm balanced on the *ludou*), and *shuatou* (head of the beam), were all in their place, and the production of bracket sets between columns was consistent with the production of sets on columns. The standardized specification in *Yingzao Fashi* listed details for various *dougong*s, and established a *cai-and-fen* system based on a *dougong*'s height to determine the size and regulation of one building.

Liao and Jin inherited the system of *dougong* from the Song Dynasty, employing a multitude of diagonal *gong*s and *ang*s, with more attention to its ornamental function.

In the Yuan Dynasty, *dougong* dwindled, employing false *ang*s with beaks becoming less common. There was a notable increase in the number of bracket sets between columns.

During the Ming Dynasty, *dougong* underwent a further reduction. The number of bracket sets between columns continued to grow, reaching as many as eight sets compared to one or two sets in the Tang and Song dynasties. The eaves were clad with a profusion of bracket sets between columns that far exceeded those on columns. This is related to the overall construction. This change aligned with the evolution of the Ming timber frame, as *dougong* lost its load-bearing functions but regained importance in terms of decoration and status. The Qing Dynasty fully inherited the Ming style, with all the beaks replaced by false *ang*s. *Gongcheng Zuofa Zeli* specified standards for *dougong* production.

To conclude, *dougong*, since its emergence in the Zhou Dynasty, evolved into an essential member of architecture in Han, matured in Tang, became standardized and characteristic of the Chinese wood frame in Song, underwent transformations in Yuan with scales dwindled, and experienced quantity reduction and production alteration between Ming and Qing when *dougong* bore no load, became smaller, gathered under eaves, and assumed a role to adorn buildings and symbolize rankings.

The structure of **dougong**

Superficially, little variations happened to *dougong* across dynasties, but in fact, both types and productions of *dougong* altered across time because of shifts in ways to build a house. Though methods to make *dougong* diversify, the basic principles remained consistent. The appellation of *dougong* was another tricky issue. Each era named *dougong* in a timely manner, and members of *dougong* got their appellations from odd and unfamiliar Chinese characters, so it is hard to remember their names. To simplify this issue, people can compare their appellations in *Yingzao Fashi* of Song and *Gongcheng Zuofa Zeli* of Qing. The fact is that, except for local discrepancies, the names of *dougong* in Song and Qing are well recognized. Usually, from era characteristics of *dougong* production, the Song system applies to what was built in Tang, Song (Five Dynasties and Jin), and Yuan, while the Qing appellations refer to Ming and Qing architecture, which conforms to timelines of architectural evolution.

Yingzao Fashi exemplified *dougong*'s structures and productions, creating a standard for Song and a basis of development for later dynasties. This Song-style *dougong* contained *dou* (bearing block), *gong* (bracket arm), *ang* (lever), *fang* (tiebeam), and *shuatou* (head of the beam). The whole structure emerged as a layer of *dou*s standing under a layer of *gong*s, which then received another layer of *dou*s on which rested another layer of *gong*s. *Dou*s and *gong*s repeatedly superposed with *gong* on *dou* and *dou* on *gong*.

Dou **and** ***sheng.*** The bottom of a bracket set was *ludou* (*zuodou* in Qing), the foundational cap block for a *zan* (called *duo* in Qing) of a bracket set, bearing the largest load. The notches on the facade were *doukou* (mortise of a cap block), which became the module unit in Qing and the scale standard to measure a building. The two sides of *doukou* served as ears of *dou* (a bearing block), under which was *douping* (a plain of *dou*) or the waist of *dou* in Qing, followed by *douqi* or the bottom of *dou* in Qing. There were two types of *douqi*: straight or concave. The ratio of *doukou*, *douping*, and *douqi* was 4:2:4, which remained unchanged across dynasties. In addition, the bottom of *ludou* in all dynasties appeared larger than *doukou*. For other parts of *dou*s, their bottoms appeared

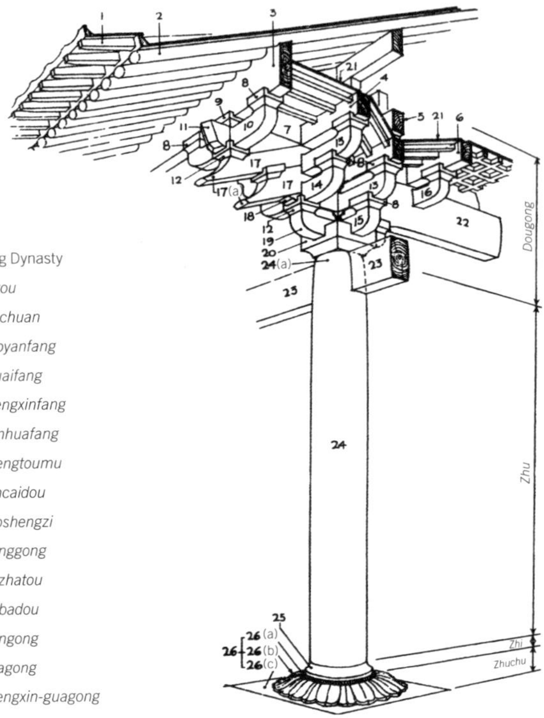

	Song Dynasty	Qing Dynasty
1	*Feichuan*	*Feitou*
2	*Yanchuan*	*Yanchuan*
3	*Liaoyanfang*	*Tiaoyanfang*
4	*Luohanfang*	*Zhuaifang*
5	*Zhutoufang*	*Zhengxinfang*
6	*Pingqifang*	*Tianhuafang*
7	*Chenfangtou*	*Chengtoumu*
8	*Sandou*	*Sancaidou*
9	*Qixidou*	*Caoshengzi*
10	*Hegong*	*Xianggong*
11	*Shuatou*	*Mazhatou*
12	*Jiaohudou*	*Shibadou*
13	*Mangong*	*Wangong*
14	*Guazigong*	*Guagong*
15	*Nidaogong*	*Zhengxin-guagong*
16	*Qifugong*	
17	*Ang*	*Ang*
17(a)	*Angzui*	*Angzui*
18	*Huatouzi*	
19	*Huagong*	*Qiaotou*
20	*Ludou*	*Zuodou/dadou*
21	*Zhechuanban*	*Gaidouban*
22	*Yanfu*	*Liang*
23	*Lan'e*	*E'fang*
24	*Zhu*	*Zhu*
24(a)	*Zhutou*	*Zhutou*
25	*Zhi*	
26	*Zhuchu*	*Zhudingshi*
26(a)	*Penchun*	*Gujing*
26(b)	*Fupeng*	*Gujing*
26(c)	*Chu*	

Each tier of the cantilever arm, either a *huagong* (19) or an *ang* (17), is called a *tiao* (jump). A set of *dougong* may be made up of from 1 to 5 *tiaos*. The example given here is one with 3 *tiaos*—1 *huagong* and 2 *angs*.

The components of *dougong* and comparisons of its appellations between Song and Qing dynasties.

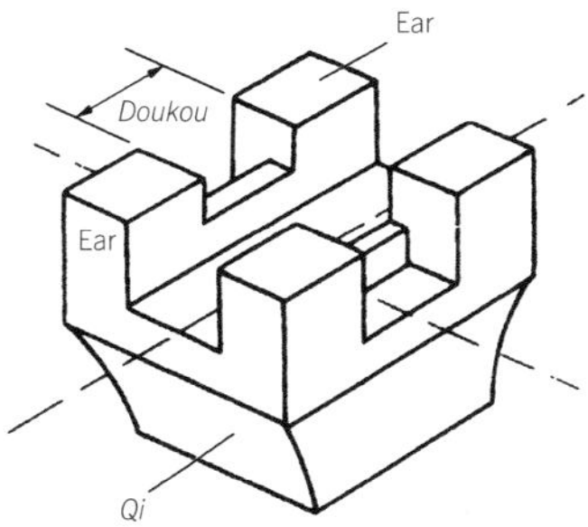

Ludou.

slightly bigger than *doukou* in Song, while the bottom was equally wide as the mortise in Qing. In Tang and before, there emerged examples of adding *minban* (a square board) beneath *ludou*, but not found in later dynasties.

Other *dou*s had *doukou* with two-sided or four-sided notches. *Sandou*s in Song (called *caoshengzi* in Qing) referred to small blocks at both ends of the horizontal arms above the *ludou*. *Qixindou* (center block) stood right on the central axis of *ludou* in a vertical direction. *Jiaohudou*s (*shibadou* in Qing) were connecting blocks extended from both sides of *huagong* (flower arm), named *qiao* in Qing. *Sandou*s (*sancaidou* in Qing) served as small blocks extended from inside and outside of two ends of horizontal arms. Generally speaking, in Qing style, *dou* was classified into *zuodou* with four-sided notches, *caoshengzi* (small block) notched on two sides, *shibadou* (connection block) notched on four sides, and *sancaisheng* (small block), and in Song, *dou* into *ludou* notched on four sides, *jiaohudou* notched on four sides, *qixindou* with notches on four sides, and *sandou* with notches on two sides. Nevertheless, they all looked indistinguishable.

In actual use, the shapes of *dou* may have been adjusted. A good instance is the main hall of Baoguo Temple in Ningbo. First of all, because bracket arms inclined at 45 degrees from two directions intersected at the corner, a bracket set on the corner under inner eaves had to flatten the lower *dou* without ears, called *pingpandou* (more of it also employed in *dougong* on the corner in Qing). Besides, as the columns in the main hall were melon-shaped and *dougong* on columns had to connect with posts under, *ludou*, therefore, was made into melon-shaped round *dou*, named accordingly, and meanwhile, the *dou*s in *dougong* between columns followed the same treatment. This type of round *ludou* and melon-

TOP *Dou*s on *dougong*, Nanchan Temple, Wutai Mountain, Shanxi Province.

BOTTOM *E'jiaodou* and *pingpandou*, Main Hall, Baoguo Temple, Ningbo City, Zhejiang Province.

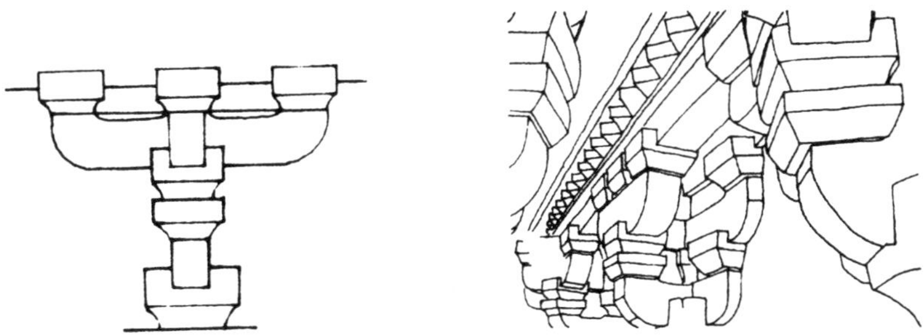

Lianzhudou, Yunyan Temple, Suzhou, Jiangsu Province.

shaped round *dou* got another name *e'jiaodou* (four corners softened into inner concave ones), because their four corners were trimmed round. Another case is in Pagoda of Yunyan Temple in Suzhou. To increase the height, one *dou* was placed on top of another *dou*, named *lianzhudou*.

Gong. *Gong*, the bracket arm of *dougong*, performed two functions. One was to complete the extension in the vertical direction of the house body, and the other was to increase the height in the parallel direction of a house to sustain the upper part. The Song system called both types *huagong* (flower arm), and Qing named them *qiao*. The shape of *gong* was a bow-shaped wood with both ends bent upward.

The shapes of *gong* could be spotted from the relics of the Han Dynasty, including rectangular, curved, zigzag, and mixed shapes. There were also animal carvings in one-big-block-and-three-small-blocks. Till Tang, the style finally got unified.[13] In Song, *dougong* members acquired specific dimensions, and a timber system was established for members such as *gong* and *ang*. This timber system was a modular one called *cai-and-fen* system. Its basic unit is *fen*, 1 *cai* (a standard-sized timber) =15 *fen*. *Chi* and *full cai* served as auxiliary units. One *chi* was 2/5 *cai* = 6 *fen*; 1 full *cai* was 1 *cai* + 1 *chi* = 21 *fen*. All rectangular components were 15 *fen* high and 10 *fen* wide, that is, height:width = 3:2. Taking *dougong* as a module meant the cross-section of *gong* was 1 *cai*, and *dou* was 15 *fen* high and 10 *fen* wide. The distance between the upper and lower *gong*s was named 1 *chi* = 6 *fen*. Full *cai* equaled 1 *cai* plus 1 *chi* = 21 *fen*.

According to *Yingzao Fashi*, *cai* in *dougong* was graded into eight classes:

First-grade *cai*: 9 *cun* high and 6 *cun* thick, used for nine- or eleven-bay palaces.

Second-grade *cai*: 8.25 *cun* high and 5.5 *cun* thick, used for five or seven-bay palaces.

Third-grade *cai*: 7.5 *cun* high and 5 *cun* thick, used for three- or five-bay palaces and seven-bay halls.

Fourth-grade *cai*: 7.2 *cun* high and 4.8 *cun* thick, used for three-bay palaces and five-bay halls.

Fifth-grade *cai*: 6.6 *cun* high and 4.4 *cun* thick, used for small three-bay palaces and three-bay halls.

Sixth-grade *cai*: six *cun* high and four *cun* thick, used for pavilions and small halls.

Seventh-grade *cai*: 5.25 *cun* high and 3.5 *cun* thick, used for small palaces or pavilions.

Eighth-grade *cai*: 4.5 *cun* high and 3 *cun* thick, used for caissons in palaces or pavilions.

The Qing standard laid its base on *doukou* of *zuodou*, which had eleven classes. The use of *dougong* as a modular unit made the scale and proportions of one building have rules to obey, achieved a proportional harmony, and met load-bearing requirements. As history evolved, the standard-sized timber in *dougong* was shrinking, which could be seen from the comparisons between the Song and Qing systems. For instance, the seven-bay main hall of Foguang Temple in Tang cost timbers of 30 cm × 20.5 cm, five-bay palaces in Song, Liao, and Jin applied timbers around 24 cm × 18 cm, but the nine-bay Hall of Supreme Harmony in the Forbidden City of Qing used only timbers of 12.6 cm × 9 cm.

*Gong*s also varied according to their locations in *dougong*. *Nidaogong* in Song or *zhengxin-guagong* in Qing (axial bracket arm) intersected with *ludou* (cap block) vertically and stood parallel to the house façade. *Gong*s that jumped out vertically were *huagong*, i.e., flower arm (in Song, *gong*s parallel to the façade were all named *huagong*; in Qing, they were named *qiao*). Above *huagong* and *ang* stood *guazigong* (oval arm), called *guagong* in Qing. At the end of all outward jumps, on the top of the last jump was *linggong* (regular arm), named *xianggong* in Qing, which no long projected upwards but stuck out in the middle

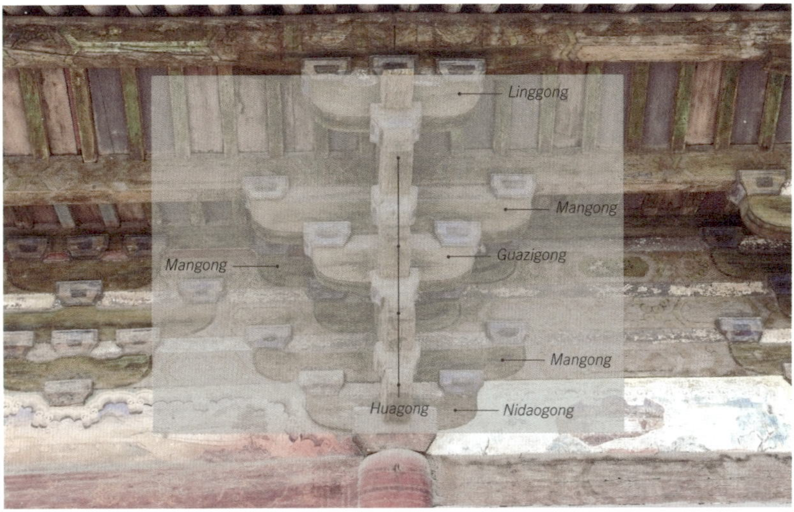

TOP *Gong*s on *dougong*, Main Gate, Dule Temple, Ji County, Tianjin.

BOTTOM *Tiao*s, East Hall, Foguang Temple, Wutai Mountain, Shanxi Province (double-*chao*-double-*ang*-seven-bracket-set).

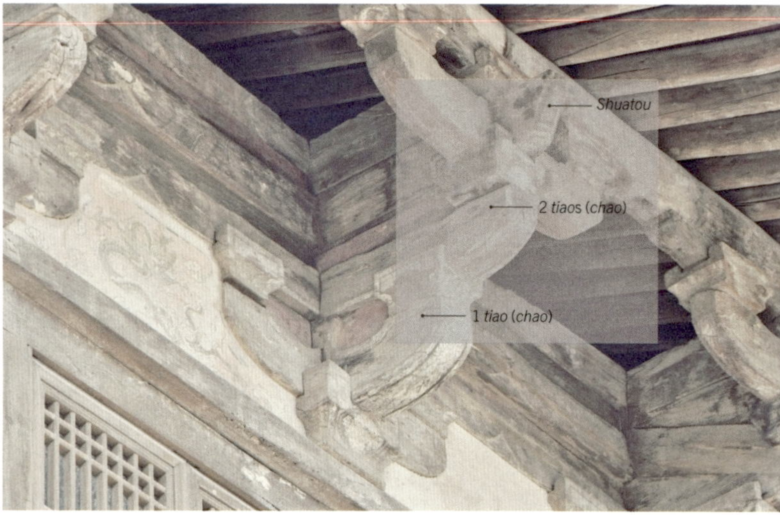

TOP *Tiao*s in double-*chao*-five-bracket-set.

BOTTOM Double-*chao*-double-*ang*-seven-bracket-set.

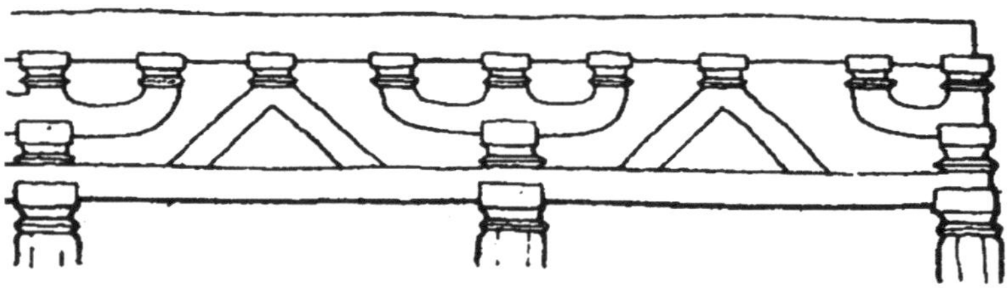

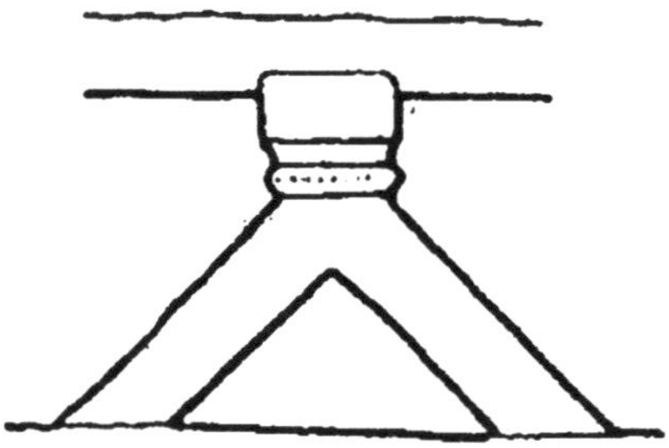

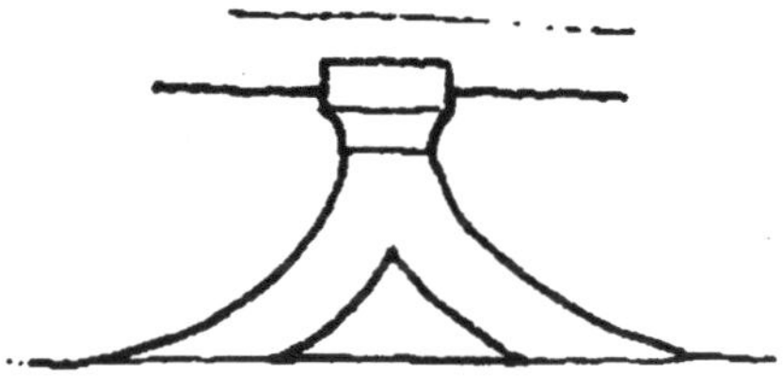

TOP Inverted-V *gong*, Cave No. 21, Yungang Grottoes, Datong City, Shanxi Province.

MIDDLE Inverted-V *gong*, Cave No. 9, Yungang Grottoes, Datong City, Shanxi Province.

BOTTOM Inverted-V *gong*, Cave No. 5, Maiji Mountain, Tianshui City, Gansu Province.

of a *shuatou* to end the whole *gong*. *Nidaogong* (*Zhengxin-guagong*), *guazigong* (*guagong*), and *linggong* (*xianggong*) were all transverse arms placed on the first tier from the bottom of *dougong*. All the other transverse arms on the second tier or above were *mangong* (long arm), named *wangong* in Qing. *Mangong* was the *gong* on top of the horizontal *gong*, which heightened the structure.

Huagong (flower arm) was divided into single-layer and double-layer. Each layer was called *chao* in Song, the single layer being one *chao* and the double layer two *chao*s. The ultimate was two *chao*s. The further outward jumps would use *ang* (a long slanted lever arm balanced on the *ludou*) instead of *gong*. One *chao* or one *ang* was also a *tiao* (jump/tier of a bracket set outward or inward); the distance of one *tiao* and the size of *doukou* determined how many *tiao*s were made. The number of *tiao*s plus three equaled the number of bracket sets. For example, two *chaos* below, single *ang* above, and three *tiao*s could be titled double-*chao*-single-*ang*-six-bracket-set. In accordance, one *tiao* made four bracket sets, two *tiao*s five bracket sets, four *tiao*s seven bracket sets, and five *tiao*s eight bracket sets. *Huagong* and *ang* were members of constructing *tiao*, but they could also produce no *tiao*.

Inverted-V *gong*. Between Han and Tang, there turned up inverted-V *gong* identical with the Chinese character "人." In early architectural materials (murals, stone carvings, engraved relief line carvings), straight inverted-V *gong* dominated between the Han and Northern Wei dynasties, while during the Jin and Southern and Northern Dynasties, curved inverted-V *gong* prevailed, either independently or in combination with a bracket set of one big block and three small blocks, or adding a short post between the feet of *gong*. Inverted-V *gong* appeared mostly in the middle of bracket sets on columns and vanished after Tang.

Diagonal *gong*. In bracket sets on corners, due to structural reasons, sometimes extended *huagong* failed to intersect with the transverse arms at right angles like in bracket sets on columns or between columns but jumped out diagonally at an angle of 60 degrees or 45 degrees. This type of *gong* was a diagonal *gong*. In Liao, Jin, and Yuan, it appeared not only in bracket sets on corners but also in intermediate bracket sets in large quantities. However, other dynasties did not find evidence of its use.[14] Therefore, the emergence of diagonal *gong*s can distinguish timber frame architecture in the Liao, Jin, and Yuan dynasties.

TOP Diagonal *gong*, Zhengjue Temple, Zhangzi County, Shanxi Province.

BOTTOM Diagonal *gong*, Sanzong Temple, Dazhonghan Village, Zhangzi County, Shanxi Province.

TOP Diagonal *gong*.

BOTTOM *Yingong*, Foguang Temple, Wutai Mountain, Shanxi Province.

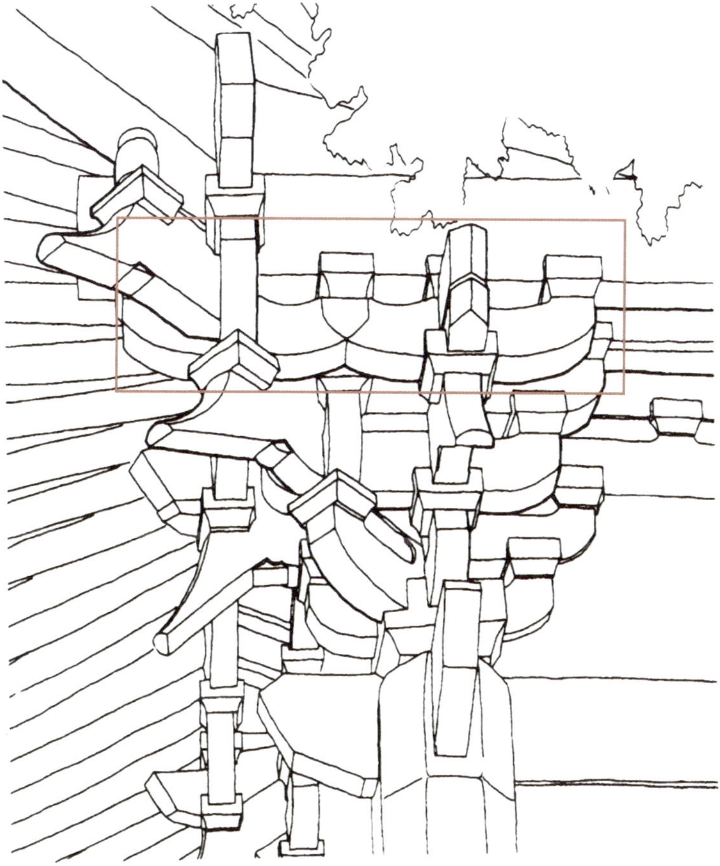

Mandarin-duck *jiaoshougong*, Main Hall, Chuzu Temple, Dengfeng City, Henan Province.

***Yinggong* (shadow arm) and *yingong* (invisible arm).** In the Tang Dynasty, colored paintings were drawn with a pattern of an inverted-V *gong* on the wall between *dougong* and columns, which were named *yinggong* (shadow arm). Another similar *gong* was *yingong* (invisible arm), on which was carved the image of *gong* on the surface of tiebeams. *Yingong* depicting *nidaogong* could be found in many buildings.

Mandarin-duck *jiaoshougong* (neck-locked arm). Strictly speaking, this *gong* belonged to *yingong*. In some cases, *dougong* on the corner was so close to the intermediate *dougong*, resulting in two overlapping horizontal *gong*s and causing a structural conflict. The solution was to carve a shared head of *gong* at the intersection, sustaining the same small block, like two mandarin ducks locking necks together, its name acquired accordingly. This *gong* had another name *liangongjiaoyin* (connecting *gong*s intersected). Their specimen could be discovered in the Xiahuayan Temple in Datong and the main hall of the Chuzu Temple of Shaolin Temple in Dengfeng, Henan.

***Dingtougong* (*chagong*, inserting *gong*).** In order to support the beams on inner eave columns and to treat structural problems at the intersection of the beam tail and the column shaft, sometimes half of a *huagong* was inserted into the column while the other half could be protruding or not according to space needs, and some protruding parts used mortise pin to prevent shedding. This form of insertion was called *dingtougong* because its shape resembled nails driven into wood. In the south, *dingtougong* was also known as *chagong*. Its specimen appears in the main hall of Baoguo Temple in Ningbo and Hualin Temple in Fuzhou.

Pengjie linggong. The way to directly support *timu* (placed in flower arm to receive eave purlin) with *linggong* (regular arm) instead of a tiebeam was *pengjie linggong*.

Dinghuamoke gong. When *gong*s coming out from the front and back of the big block that sustained the ridge purlin collided with the inverted V-shaped brace that fixed the ridge purlin, they were chopped off diagonally and connected with the brace through mortise and tenon.

Accounted heart and stolen heart. According to *Yingzao Fashi*, there were two techniques to make bracket sets. The regular one was to install one horizontal *gong* (arm) on every *tiao* (jump/tier of a bracket set outward or inward)

TOP *Dingtougong*, Main Hall, Baoguo Temple, Ningbo City, Zhejiang Province.

BOTTOM *Pengjie linggong*, Tiantai Temple, Pingshun County, Shanxi Province.

TOP *Dinghuamoke gong*, Kaishan Temple, Gaobeidian City, Hebei Province.

BOTTOM Stolen heart and accounted heart, Main Gate, Dule Temple, Ji County, Tianjin.

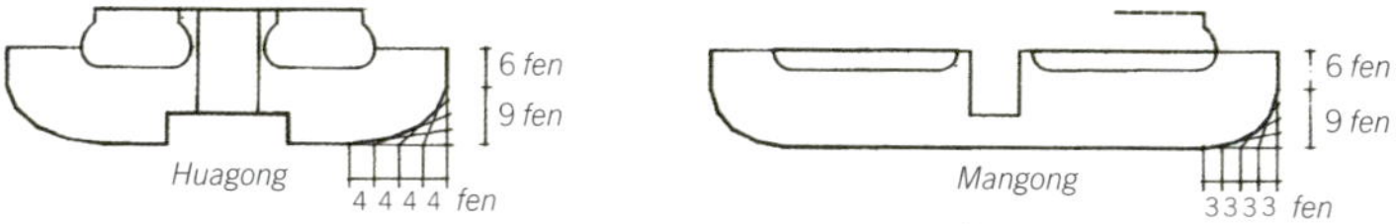

TOP Ways to make *gong* head entasis in *Yingzao Fashi*.

BOTTOM Three concave lobes on the gong head of the Main Hall of Nanchan Temple, Wutai Mountain, Shanxi Province.

or beak, called the "accounted heart." The other simplified technique omitted that horizontal *gong* (arm), therefore named the "stolen heart." Both ways were employed across dynasties, with "stolen heart" more popular in Tang and Song while "accounted heart" with double *gong*s prevalent after Jin and Yuan.

***Gong* head entasis.** At both ends of *gong* were upward-curved outer arcs. This method was called entasis or rounded end. In the Han Dynasty, *gong* head entasis had such forms as curves and broken lines. During the Southern and Northern Dynasties, it was divided into multiple lobes, each with concave arcs and an obvious style. In Tang, its forms of expression varied. The *gong* head of the main hall of Nanchan Temple had three concave lobes, while the split lobes of Foguang Temple looked vague and more rounded. In *Yingzao Fashi*, the *gong* head entasis was all made of broken lines, with five lobes in *linggong* (regular arm) and four in other *gong*s, but methods differed in practical situations. In Ming and Qing, the *gong* head entasis turned smoother and fuller.

Ang. As a slanted and once the largest member in *dougong*, it functioned basically as a lever. Its front end was to raise and support the overhanging eaves while its tail was held down by the beam, and with the rear pushing the front to rise, *ludou* lying in the middle constituted the fulcrum of the lever. *Ang* could be divided into *shang'ang* (beak) and *xia'ang* (tail). The tail was more common, while *shang'ang* was exclusive to indoors, in a balcony *dougong*, or over *tiao* (jump/tier of a bracket set outward or inward) inside a *dougong*.

Before the Tang Dynasty, *ang* was invisible among the physical remains or related image materials. The earliest example was split bamboo *ang* (with a straight-bevel beak) in the bracket set on columns in the main hall of Foguang Temple. *Ang* used in Tang was larger than in the later dynasties, and the largest *dougong* in history was constructed. Both single and double *ang*s were used. By Song, double *ang*s gained more popularity. In Yuan, false *ang*s appeared in the shape of a beak on the outward turn and as more ornamental than levering. The combination of real and false *ang*s also existed, for example, real *ang* in a bracket set on columns and false *ang* in the intermediate set. Ming Dynasty continued this combination. As time evolved, *ang*s gradually shrank, and their functions mutated. In Ming and Qing, all the rear parts of the extension of *huagong*, *ang*, and *shuatou* were folded and extended to the bottom of purlins, which was called *liujin dougong*, and the structural functions of a slanted *ang* disappeared. In a

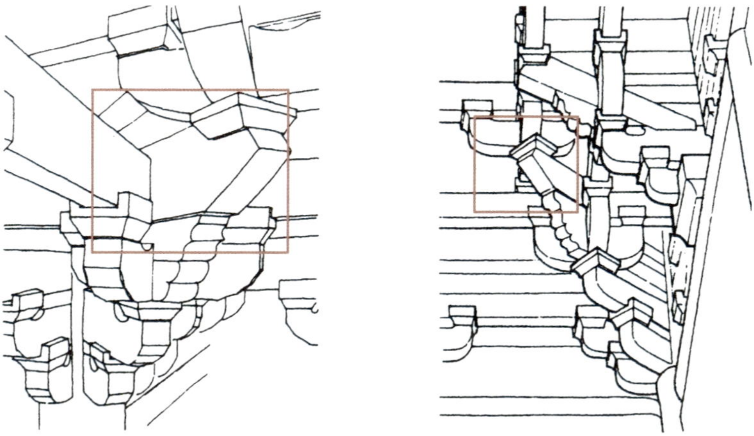

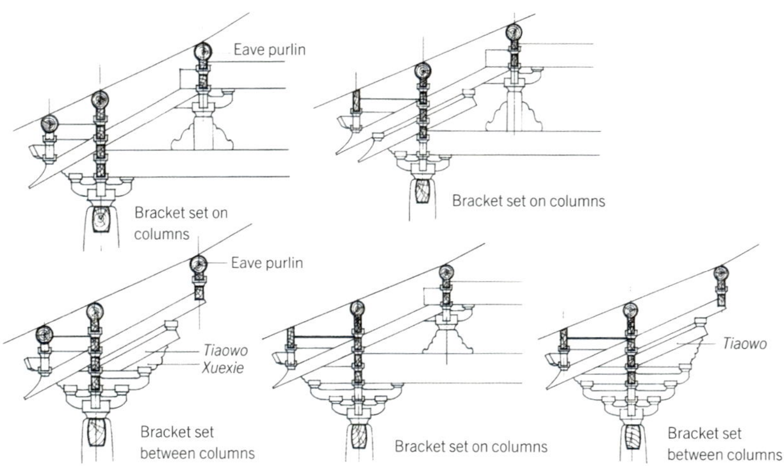

TOP *Shang'ang* inside Sanqing Hall, Xuanmiao Temple, Suzhou (not commonly used).

BOTTOM Some ways to use *ang*.
Note: Tiaowo means support component and *Xuexie* means wedge-shaped padding.

bracket set on the corner, no evidence of *ang* was displayed in Tang, and both real and false *ang*s came into use from Song to Qing.

The shape of *ang* was distinctively era-matching. Tang style preferred flat-headed *ang* with vertical truncation and split bamboo *ang* with straight-bevel beak, both of which signified simplicity and strength. Song Dynasty witnessed split bamboo *ang* and lute-face *ang* with pulvinated concave-bevel beaks. In Yuan, lute-face *ang* remained to be used, but the concave bevel was deeper than in Song, so the bottom of beak was slightly upturned and the beak was shortened and thickened. Ming-style lute-face *ang* presented an even shorter and thicker beak, and the bevel curve turned gentle without clear-cutting fold lines. This

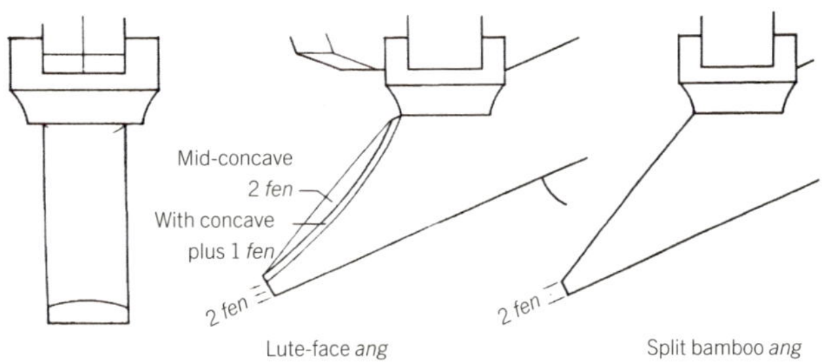

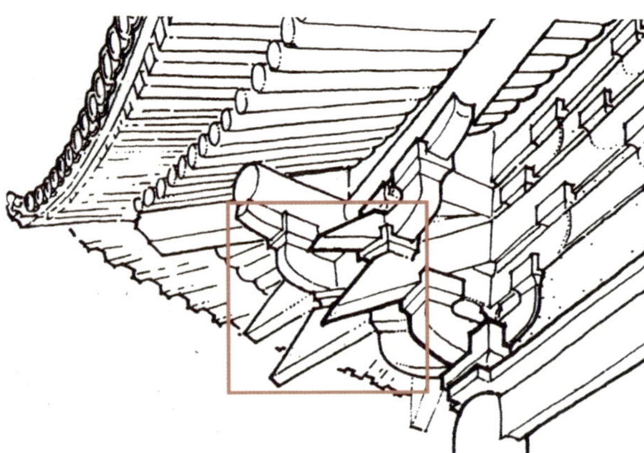

TOP Ways to make split bamboo *ang* and lute-face *ang* in *Yingzao Fashi*.

BOTTOM Split bamboo *ang*, Yuhua Hall, Yongshou Temple, Yuci District, Jinzhong, Shanxi Province.

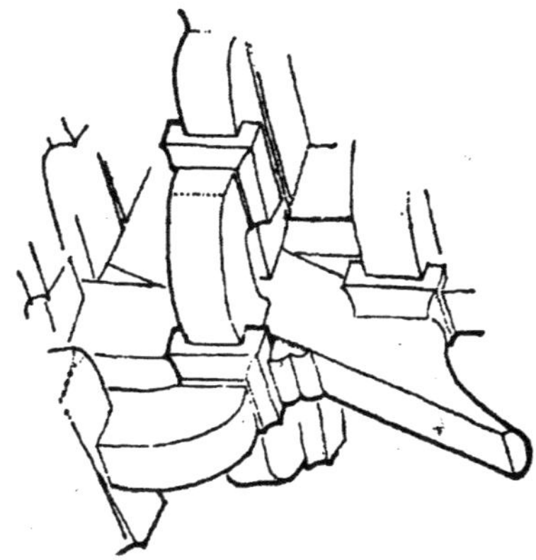

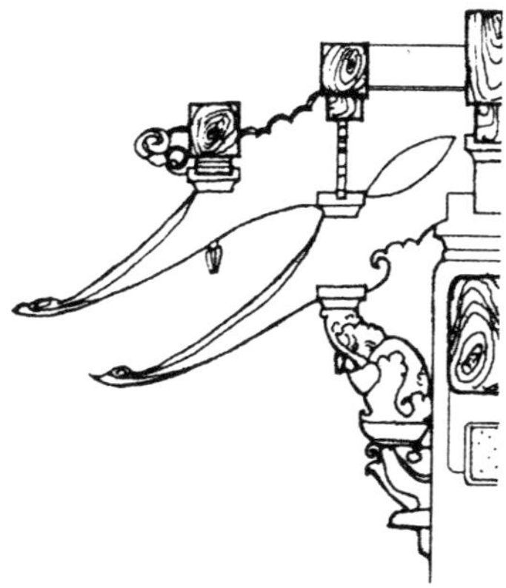

LEFT Lute-face *ang*, Shengmu Hall, Jinci Temple Jinci Temple, Taiyuan City, Shanxi Province.

RIGHT Phoenix-head *ang*, Ye Ancestral Hall, Xukou Village, Yongjia County, Zhejiang Province.

Ming style continued its influence into mid-Qing. In the late Qing Dynasty, elephant-trunk *ang* stayed popular, and at the end of Qing, hollow-*carved ang* emerged while phoenix-head *ang* entered folk houses. The earliest elephant-trunk *ang* left its trail in Yuan and Ming with its beak-like curled trunk and four bevels on the upper and lower sides of the concave bevel.

False *Ang*. *Ang* not only acted as a lever to bear the load in the building structure but also displayed beauty and elegance. In the development of architecture, to achieve a unified visual beauty of *dougong*, or to conform to the aesthetic understanding of *dougong*, false *ang*s were created for their ornamental effect instead of their structural functions. A popular false *ang* was made of out-turned-*huagong*, applied more in Ming and Qing than in Song and Yuan. A specimen was discovered in the outer eave *dougong* of the Sanqing Hall of Xuanmiao Temple in Suzhou. There were also two other forms of false *ang*s: *youang* and *cha* (inserting) *ang*. *Youang* refers to *shuatou* made into the shape of *ang* and unified with *xia'ang*, employed on bracket sets on corners. An example could be seen in the outer eave bracket set around the corner of the main hall of Qinglian Temple in Jincheng. *Cha'ang*, with beak only, was directly inserted diagonally under a block. Specimens were located in the Sansheng Hall of Shanhua Temple in Datong and the Chuzu Temple of Shaolin Temple in Dengfeng.

TOP Split bamboo *ang*, East Hall, Foguang Temple, Wutai Mountain, Shanxi Province.

MIDDLE Split bamboo *ang*, Shengmu Hall, Jinci Temple, Shanxi Province.

BOTTOM Qing elephant-trunk *ang*.

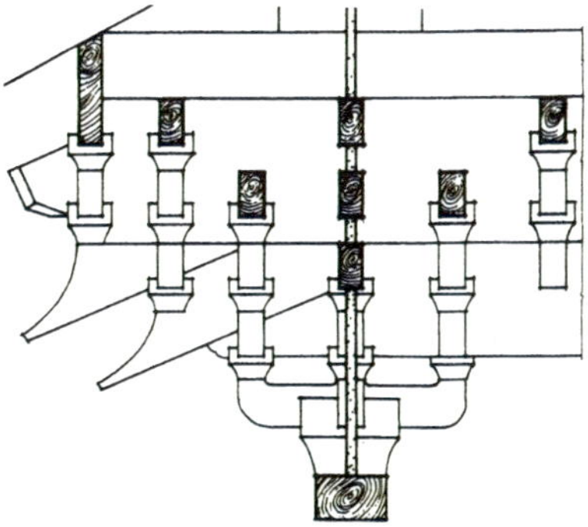

TOP LEFT *Cha'ang* (false *ang*), Sansheng Hall, Shanhua Temple, Datong City, Shanxi Province.

TOP RIGHT, ABOVE MIDDLE, ABOVE LEFT, ABOVE RIGHT False *ang*.

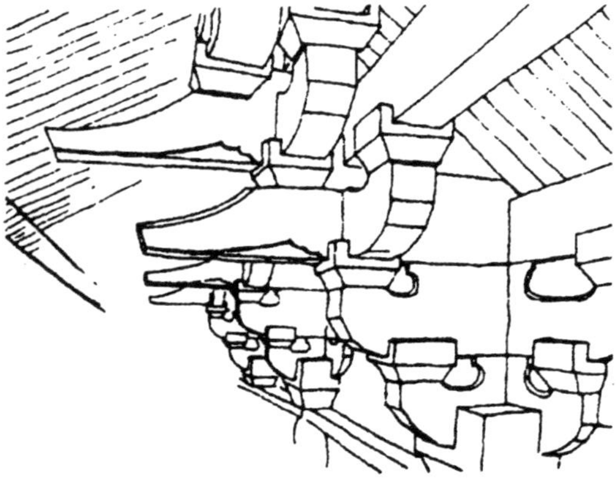

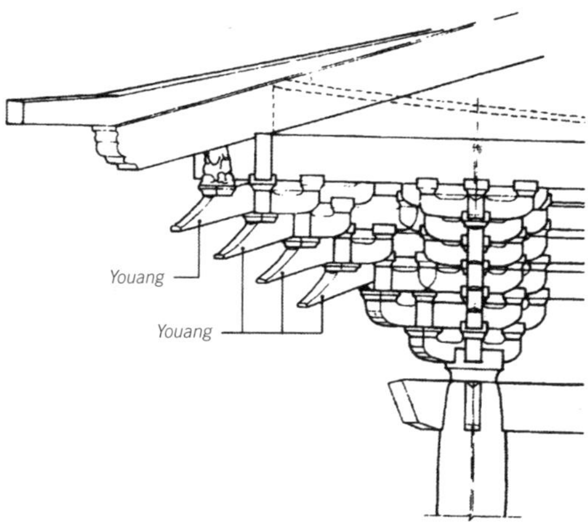

TOP Outer end of *huagong* made into the shape of *ang* (false *ang*), Sanqing Hall, Xuanmiao Temple, Suzhou.

BOTTOM *Shuatou* made into the shape of *ang* (false *ang*).

***Shuatou* (head of the beam).** It was the horizontally protruding component of the uppermost layer of *dougong*. Generally, the head of the beam was stretched out to make a *shuatou*. The existing examples indicated its appearance after Tang. Its diverse forms included a vertically truncated head, split bamboo head, rolled-petal head, the Song-standard grasshopper head, and many other special-shaped heads. On the whole, the vertically truncated and split bamboo heads were employed in Tang, Song, Liao, and Jin, but not in Yuan, Ming, and Qing. The size of *shuatou* was shorter in Ming and Qing.

***Fang* (tiebeam).** A horizontal tiebeam was placed on each *tiao* of *dougong*. This beam connected each set of *dougong* into a whole, making the scattered *dougong*s stronger under the combined force of the whole structure. Tiebeams on the axis of the column capital, i.e., axial bracket arm, were called *zhutoufang*, those on the oval arm or long arm named *luohanfang*, those on the regular arm jumping outward were titled eave purlins, and those on the regular arm jumping inward were known as *pingqifang*.

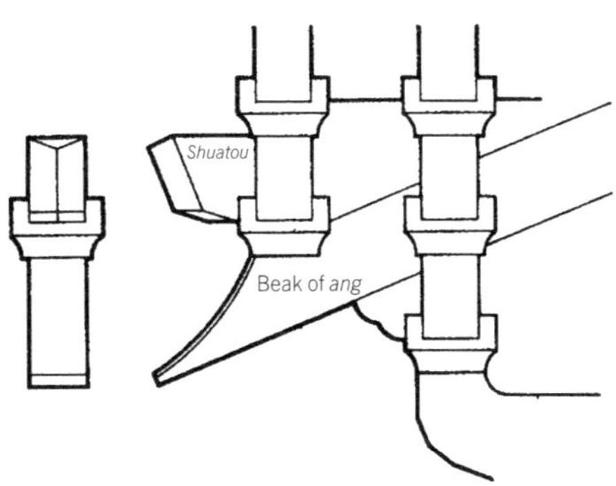

Beak *of ang* and *shuatou*.

TOP *Shuatou*, Kaishan Temple, Gaobeidian City, Hebei Province.

BOTTOM *Shuatou*, Dacheng Hall, Confucian Temple, Zhengding County, Hebei Province.

TOP *Shuatou*, Jinci Temple, Taiyuan, Shanxi Province.

MIDDLE *Shuatou*, Nanchan Temple, Wutai Mountain, Shanxi Province.

BOTTOM *Shuatou*, Yanqing Temple, Wutai Mountain, Shanxi Province.

FOLLOWING PAGE *Shuatou*, *chao*, *ang*, East Hall, Foguang Temple, Wutai Mountain, Shanxi Province.

Chao

Shuatou
Ang

Evolution of the *shuatou*.

Beams, Purlins, and Rafters

Chinese wooden architecture is a framing structure whose key members are beams, purlins, rafters, and columns.

The evolution of beams, purlins, and rafters

We know from archaeology that ancient Chinese architecture employed both timber frame and load-bearing walls ever since primitive society.[15] Many residential houses excavated at the Banpo site of Yangshao culture in Shanxi had small superficies, roughly equivalent to that of a current house, and shaped round, oval, or rounded square in the plan. In the center of a house stood several columns to support the roof, which was made of closely arranged rafters and coated with a layer of grass mud, creating an umbrella shape. The walls were constructed by snugly arrayed wooden pillars wrapped with thick grass mud. These walls demonstrated features of both wood-frame and load-bearing walls. This type of wall could be seen in the Yangshao culture Zhengzhou Dahe Village site and the Palace site of the early Shang Dynasty in Yanshi, Henan.

Later, in the structural development, the dense stand columns in walls gradually turned into regular rows of columns. These columns served as the skeleton of walls and later sustained the roof structure independently, which constituted the prototype of column and tie construction. This transformation evolved the column frame into a beam frame, and a system of Chinese-style timber frame structures was officially established.[16] The rows of columns in walls created a column network. In the early days, this network was aligned with the roof structure. Columns were connected by beams to maintain stability, and a whole frame was fixed to enhance holistic firmness. When more space became necessary inside a house, these columns caused problems. The solution was to eliminate columns by replacing connecting beams with load-bearing large beams, which led to the birth of a beam frame of Chinese wood construction.

Beams, purlins, and rafters, Zhengjue Temple, Zhangzi County, Shanxi Province.

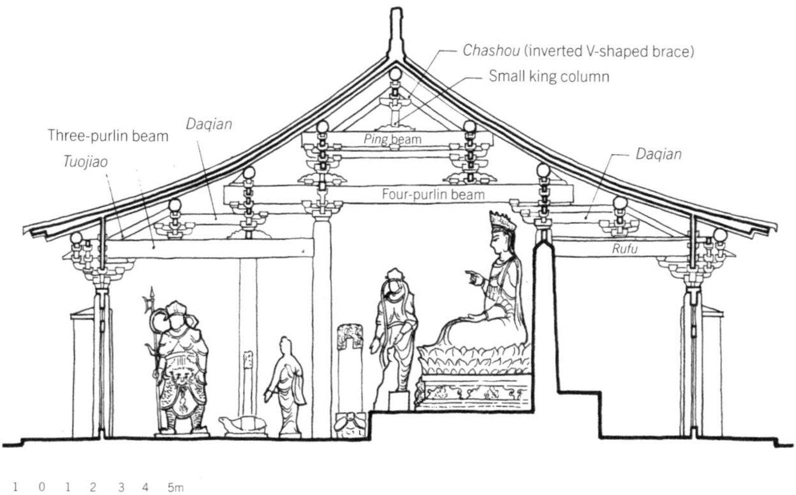

Components of a beam frame.

The composition of beams, purlins, and rafters

Beam. It was a large beam supported on a column in the width of a building, called *fu* in Song, and *liang* in Qing. According to their location and load-bearing capacity, beams could be divided into main beams and secondary beams. The main beams were erected between the front and rear eave columns in width, spanning most or all of the interior space, and dwindling beams were put above to construct a sloping roof. The beams made in Song were named after the number of rafter sections separated by purlins spanned by each tier of the beam, while the beams of Qing were called after the number of purlins spanned by each tier of the beam. For example, for a beam spanning sixrafters, Song system called it six-rafter beam but Qing seven-purlin beam; for a beam spanning four rafters, Song named it four-rafter beam but Qing five-purlin beam, and so on. A beam with one rafter was titled *daqian* in Song and *baotou* beam or one-step cross beam in Qing; the two-rafter beam was also known as *rufu* or two-step cross beam in Qing. But the topmost two-rafter beam had another name *pingfu* in Song and the top beam in Qing.

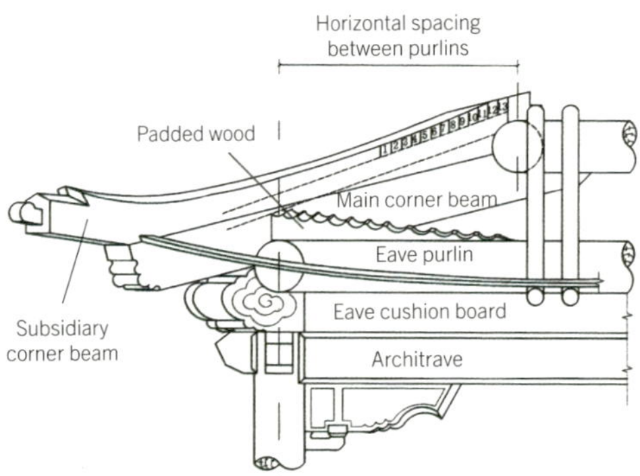

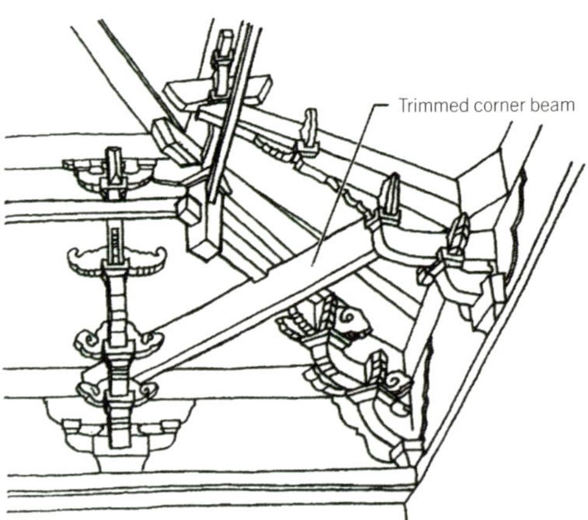

TOP Four-rafter beam and *ping* beam, Tiantai Convent, Pingshun County, Shanxi Province.

MIDDLE Main corner beam and subsidiary corner beam.

BOTTOM Trimmed corner beam.

At the corner of the eave, a beam with an angle of 45 degrees protruded as a hip rafter (called *yangma* in Song). It was made of two superposed beams, the upper one being subsidiary corner beam and the lower one main corner beam.[17] There were also trimmed corner beams at a right angle to the hip rafter on the plane and stepped corner beams in the same direction as the hip rafter.

In appearance, beams could be either straight or crescent. Crescent beams, also named rainbow beams in Han, had arc-shaped shoulders with a concave bottom and decorated sides with a lute face and low relief. They emerged in residential buildings across regions with names of unique local features. For example, in ancient Huizhou areas such as southern Anhui, western Zhejiang, and eastern Jiangxi, a crescent beam got the name of a winter-melon beam because of a similar shape; in Fujian province, an oblique beam with one end high and the other end low, was vividly depicted as a shrimp beam or a cat beam.

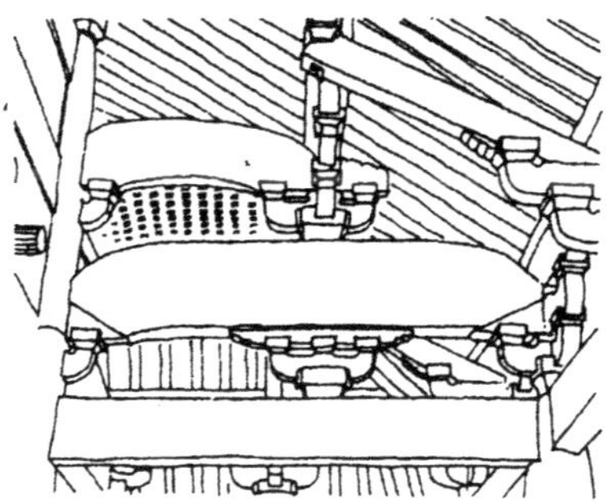

TOP Crescent beam, Er'shan Gate, Tiger Hill, Suzhou, Jiangsu Province.

BOTTOM Winter-melon beam, Huizhou City, Anhui Province.

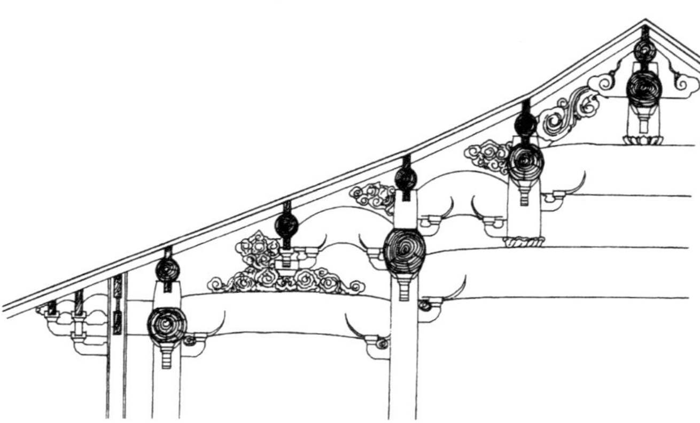

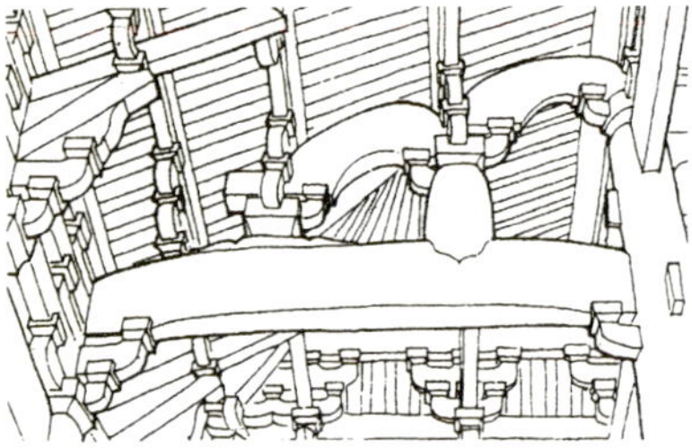

TOP Shrimp beam, Main Hall, Yanfu Temple, Wuyi County, Zhejiang Province.

BOTTOM *Caofu.*

Another kind of beam was *caofu*, a rough beam concealed above the caisson ceiling. However, in practice, this rough beam, or even slightly straightened logs, was widely applied to some ancient wooden buildings in Shanxi, especially to residential houses. In certain cases, because of limitations of certain beams, such as shortness in length, two pieces of wood would be butt-joined at the column head into one beam, or when part of wood was too thin to be fit for a beam, an inlay method would thicken that part. Besides, there was a special kind of padded wood made to expand the cross-section of a beam to strengthen its stress-bearing ability, which was named *jiaobei*.

The cross-sections of beams were shaped mostly rectangular (some were round, and even logs were used directly) with a ratio of 3:2 between height and width in Song, but nearly square in Ming and Qing. In the early eras, beam heads were cut vertically. In Tang, split bamboo heads prevailed, but in Song and Yuan, grasshopper heads enjoyed popularity, and cluster cloud or peach tip heads dominated in Ming and Qing. Beam connections diversified, either inserting the beam head directly into *dougong* (a bracket set) on the column capital and protruding to be the head of the *dougong* (regarded as a key feature of the Tang timber frame) or resting on a cap block of the column capital, or the end of the beam penetrated into the column shaft with the lower part sustained by inserting *gong*; or the back of the lower beam receiving king post, camel-hump, and *dougong*.

Purlins. They were connected with components in intersecting directions and then received rafters and undertook roofs, named *tuan* in Song and *lin* or *heng* in Qing. According to their positions, purlins fell into categories of ridge purlin (named *jituan* in Song or *jiheng* in Qing), upper purlin (called *shangpingtuan* in Song or *shangjinheng* in Qing), intermediate purlin (titled *zhongpingtuan* in Song or *Zhongjinheng* in Qing), and lower purlin (called *xiapingtuan* in Song or *xiajinheng* in Qing), axial purlin (named *beef-spine tuan* in Song or *zhengxinheng* in Qing), and eave purlin (titled *liaofengtuan* in Song or *tiaoyanheng* in Qing).

In order to strengthen the beam frame, tiebeams stood under purlins or parallel to purlins for a firmer connection. This tiebeam was named *tie to a purlin*, and in Qing, according to its respective positions, it was titled ridge tiebeam, upper purlin tiebeam, lower purlin tiebeam, and eave tiebeam. In the Song Dy-

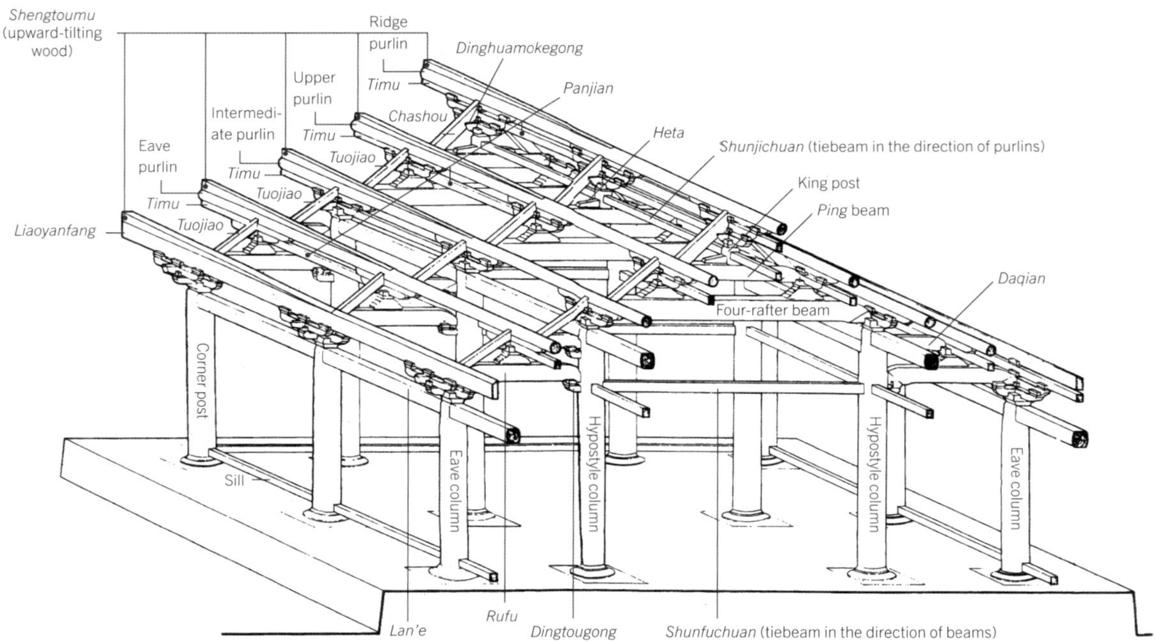

Beam frame of a building.

nasty, *timu*, its ends applying arc-shaped entasis, stood as an auxiliary to where *dougong* got connected to a purlin. Another supporting tie lay under *timu* to receive *dougongs*, named *panjian*. *Panjian* was as high and thick as the connected *dougong* and as long as a bay, with both ends inserted into the half-bracket arm. On the upper back of the eave purlin was padded with triangle-shaped wood to raise the roof vertically on a curved surface.

Purlins often extended over the two gables of a house to give them protection. This kind of extension was called *chuji*. The width of a house determined how long the extension could be. *Yingzao Fashi* regulated that a two-rafter house had a protrusion of 2 to 2.5 *cun*, and a four-rafter house had a protrusion of 3 to 3.5 *cun*.

Rafters. Rafters, whose cross sections looked round or square or half round, lay on purlins to support the roof sheathing (or roof bricks) and intersected with purlins. Their appellations varied with their positions. On cornices were eave rafters, and on roofs ridge were rafters. On a round ridge roof, curved rafters lay at the top, also called *luoguo* rafters. These rafters also appeared in small buildings such as pavilions and windowed verandas. To enhance the protrusion and rise of eaves, flying rafters stood above eave rafters.

The early evidence of eave rafters emerged under the eaves of Han stone tomb shrines and paired gate piers, such as Gaoyi *Que*, Ya'an, Sichuan Province, and the Yicihui Stone Column, Dingxing, Hebei in Northern Qi Dynasty. A specimen in wooden structures turned up in Foguang Temple where square-shaped flying rafters rested above round-shaped eave rafters. The conclusion could be drawn that applications of flying and eave rafters were fixated in the Tang Dynasty.

Every rafter had a connecting line. In Tang and Song, the upper and lower rafter heads were staggered and later developed into the oblique butt joint of the two rafter heads. A single board stood between rafters to fix each rafter, called a rafter partition board, and on higher-end buildings, a rafter bowl board functioned similarly, with holes as big as the rafter diameter dug into the board.

Other components in beam structures

King Post. It had other names, such as dwarf post or melon post. In the early eras, the king post stood under the ridge purlin, while camel-humps, *dougong* or short posts supported other parts of the beam structure. After Yuan, king posts started to be used, mostly circular in cross-section, with a few being square. A specimen of a square king post is located in the Main Hall of Nanchan Temple.

***Chashou* and *Tuojiao*.** *Chashou* was a diagonal bracing that strutted the ridge purlin and king post. Under the ridge purlin, it usually cooperated with king posts but sometimes worked individually, as can be seen in Foguang Temple. *Chashou* was popular in Song, Liao, and Jin but became thin and long in Yuan and disappeared in Ming and Qing, especially in high-ranking buildings. *Tuojiao* was akin to *chashou*, but differed in their positions. *Chashou* stood under the ridge purlin while for *tuojiao*, its bottom end rested on a beam, and the upper end held the side of a purlin. They emerged at the earliest in Tang architecture as the main hall of Nanchan Temple and the east hall of Foguang Temple. *Tuojiao* prevailed in Song, Liao, and Jin and appeared in different buildings in Yuan, but it nearly vanished in Ming and Qing.

Camel-hump and *heta*. A camel-hump was a common structural component shaped like two humps of a camel positioned under *dougong* (a bracket set) or king post to support the beam firmly, used across dynasties with variations in scale and shape. There were two types of camel-humps: full or half. A rare example of a half camel-hump was spotted in the main hall of Foguang Temple where the bottom of *huagong* (in Song, *gong*s parallel to the façade were all named *huagong*) and a tiebeam extended, and a half camel-hump stood to carry a connection bracket and a regular arm. The most common Song humps showed up in concave or convex shapes with tail entasis along with other patterns of olecranon, concave, or hat. During Liao and Jin, aside from inheriting the previous forms, new variations emerged. The camel-hump of the Front Hall of Jinci Temple reached 70 cm high. Yuan humps turned simpler, abandoning the concave and convex forms, while Ming and Qing preferred cloud patterns or lotus leaf piers.[18] *Heta* flanked the foot of the king post to fasten the king post and the beam under. It looked like a hump but different in shape and protrusion. *Heta* could be shaped rectangular or arc, or like a broken line or double-concave-olecranon-hump, the last of which was displayed in the Wenshu Hall of Foguang Temple. Furthermore, a camel hump was observed on the exterior of the king post, while the *heta* was located within the boundary.

Components of a beam structure, Front Hall, Jinci Temple, Shanxi Province.

TOP *Heta*, Fotou Temple, Pingshun County, Shanxi Province.

MIDDLE *Chashou*, *tuojiao*, and camel-hump, Yuhuang Temple, Bu Village, Zhangzi County, Shanxi Province.

BOTTOM King post.

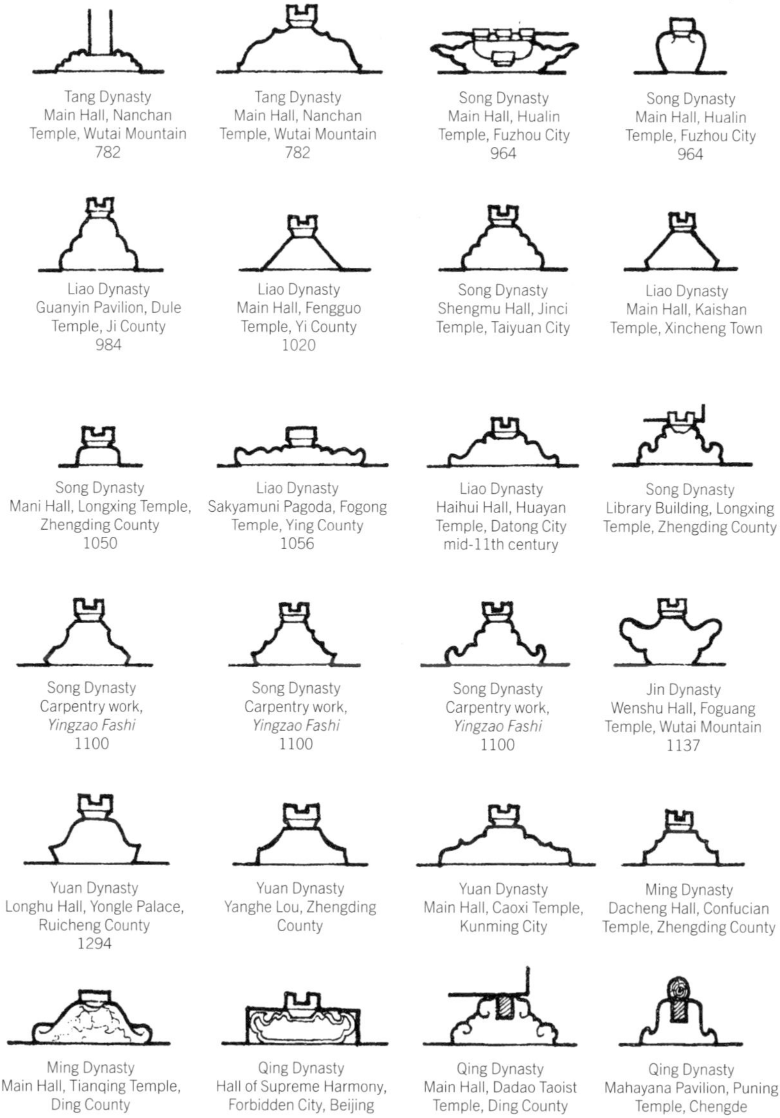

Evolution of camel-humps.

Bays, rafters, and house scale

The scale of an ancient building was typically gauged by the number of bays in length or the number of rafters in width. It was challenging to ascertain the origin of this measurement. In AD 827, during the early years of Emperor Tang Wenzong, an edict was issued stipulating that "the residences of princes and nobles should not be equipped with double *gong*s and caisson ceilings, and the three-grade halls have five bays and nine purlins." Therefore, it can be reasonably assumed that the use of bays, rafters, and purlins as house measurement scales was already prevalent during the mid-Tang Dynasty. It is likely that these scales were in use at least before the mid-Tang period.

In the traditional Chinese language, the term "bay" has two distinct meanings. The first explanation pertains to the number of houses in "with tens of thousands of houses" (from Tang Du Fu's poem) or space in general in daily expressions like "one bedroom." The other sense describes a concept of scale in the number of bays in length, which stipulates the size of a house. In the latter sense, a bay was defined as the space between two front and two rear columns or between two beam frames that contained beams and the front and rear eave columns in width. In the north, this beam frame was known as *feng*, while in the south, it was called *tie*. The number of beam frames constructed was directly proportional to the number of bays. The distance between the two frames constituted the length of a bay, while the distance between the two purlins in width (the projected horizontal length of rafters) produced the length of a rafter. Consequently, the length of a bay and the width of the rafters can be used to accurately depict the scale of a building.

In fact, this scale measurement was determined by the construction of structures. The elevation of a house made it possible to distinguish the beam frames that bore load. This enabled the columns under the frames to divide the front of a house into regular intervals, which were then used to determine the bay length. The bay length was subsequently adopted as the standard unit of measurement for the length of a house. However, the distance between columns could not be used to measure the width of a house because, except for eave columns, the intermediate pillars were often increased or decreased according to the length of the beams. Furthermore, columns were often reduced when space was limited, so the length and number of rafters offered a better way to measure width.

Rafter amounts can be used to scale a building since it can refer to the depth of a building. In this way, the building is typically called an X-rafter house. Some buildings have a number of rafters that are equal to the length of the longest beam, while others include the lengths of multiple beams. For example, if a building has a depth of six rafter, it's also called a "six-rafter house." Specifically, for a "six-rafter house," the eaves can be supported by six beams and two columns at the ends.

Four-rafter beam (four-rafter house).

Four-rafter beam plus *rufu* (six-rafter house).

Three-rafter beam plus *daqian* (four-rafter house).

Five-rafter beam plus rear *daqian* (six-rafter house).

NOTES

1. Chen Mingda, *Ancient Chinese Wooden Structure Construction Technology (Warring States—Northern Song Dynasty)* (Beijing: Cultural Relics Press, 1990), 10.
2. Ding Yuan, *One Book to Understand Chinese Architecture* (Beijing: Zhonghua Book Company, 2012), 267.
3. Fu Xinian, *An Overview of Ancient Chinese Architecture* (Beijing: Beijing Press, 2016), 83.
4. Guo Huayu, *The Origin of Chinese Classical Architectural Forms* (Wuhan: Hubei Education Press, 2015), 181.
5. Ibid., 179.
6. Zhang Jiaji, *Discussions on Chinese Architecture* (Taiyuan: Shanxi People's Publishing House, 2003), 399.
7. Liu Dunzhen, *The Complete Works of Liu Dunzhen*, vol. 6 (Beijing: China Architecture & Building Press, 2007), 186.
8. Li Jinlong, *Identifying Chinese and Foreign Ancient Buildings* (Shanghai: Shanghai Bookstore Press, 2016), 62.
9. Li Yunhe, *Principles of Chinese Classical Architectural Design* (Tianjin: Tianjin University Press, 2005), 99.
10. In his book *The History of Ancient Chinese Architecture*, Mr. Liu Dunzhen cites the structural description of the architecture of the *ling-gui* in Mr. Chen Mengjia's article "Western Zhou Bronzes: A Breakdown of the Generation" in the 10th volume of the *Journal of Archaeology* of December 1955 and argues that the shapes and combinations of these elements are roughly the same as those used in the construction of the gables and columns of the later generations. Later, Mr. Pan Guxi of Southeast University in *The History of Chinese Architecture* and Mr. Guo Huayu in *The Origin of Chinese Classical Architectural Forms* both gave such a description.
11. Pan Guxi, *The History of Chinese Architecture* (Beijing: China Architecture & Building Press, 2015), 277.
12. This conclusion was reached by Mr. Tang Dahua from Dezhou, Shandong, through a comprehensive examination of a multitude of examples presented in his online publication.
13. Pan Guxi, *The History of Chinese Architecture* (Beijing: China Architecture & Building Press, 2015), 279.
14. In 2013, the large tomb of the Northern Dynasty in Jiuyuanggang, Xiashe Village, Xinzhou City, Shanxi Province was excavated. A mural was discovered above the tomb door. The painting depicted the Youxiang Hall building with a high degree of realism. The diagonal *gong* was distinctly applied to the bracket set on columns, but there are also beliefs that it was the result of the artist's drawing perspective.
15. Until today, the dual structure remains evident in certain residential edifices.

16. Li Yunhe, *Principles of Chinese Classical Architectural Design* (Tianjin: Tianjin University Press, 2005), 203.
17. Liu Dunzhen, *The Complete Works of Liu Dunzhen*, vol. 6 (Beijing: China Architecture & Building Press, 2007), 196.
18. Ibid., 202.

CHAPTER 5

FACADE

Visual Transformation of the Facade

The facade, conventionally recognized and, in a sense, true, embodies the main external appearance of the entire building, serving as the key element for visual focus and aesthetic appeal. It is the first thing that catches the eye because of the prominence of its volume. This notion has indeed been consolidated in Western architecture or modern architecture, where the variation of the building's form depends primarily on the structural changes of the facade. The morphology, lines, and even materials of the facade are directly related to the overall form and visual perception of the building.

The visual experience and perception we attain in modern architecture, however, cannot be replicated in traditional Chinese architecture. Whether it is a collective or individual dwelling, the facade of ancient buildings does not occupy the visual center in terms of the overall appearance of the structure. Even in terms of its own form and visual perception, the visual impact of the facade is minimal. This is particularly evident when observing ancient Chinese architecture from a medium to long distance. In general, individual dwellings in traditional Chinese architecture do not exist in isolation; they usually appear in the form of courtyard clusters. In modern times, the remaining individual ancient buildings are mostly surrounded by collapsed or damaged structures that cannot be preserved intact. When observing ancient architecture from a distance, the visual perception is one of collectivity, where the dominant visual element is the grand roof of the buildings. This is also the reason why ancient Chinese architecture highlights the prominent feature of the grand roof. The poetic descriptions such as "surrounded by mountains and trees, with overlapping eaves like a hundred steps" and "resembling wings, resembling thorns, like a bird's skin, like soaring flight" reflect the visual impression created by the overlapping roofs of the buildings. The facade, when viewed from the front of the ancient architectural structures, apart from the elevated platform, the grand roof, and the layered *dougong*, occupies a minimal visual proportion. They often recede below the eaves, creating a space of subdued presence. As a result, when viewing the dwelling from a distance, the facade naturally takes on a secondary visual impression.

The distant view of ancient Chinese architecture.

Due to the collective nature of ancient architectural dwellings, it is often challenging to achieve a medium-distance viewing experience. Within the courtyard layout, a close-up visual experience is created, and the facade of the buildings undergoes a transformation depending on the viewing distance. Traditional courtyards are enclosed by surrounding buildings that form the four walls of the courtyard space. At this point, the facades of the enclosed buildings become the four walls of the courtyard space, and the exterior of the dwelling's facade becomes the interior; in fact, this transformation not only creates an outward-facing facade but also draws attention to the interior experience of the courtyard space. When the inward-facing nature of the exterior facade is emphasized, there are changes in the style and form. First, the image of the exterior facade of the buildings surrounding the courtyard should be in harmony with the atmosphere of the courtyard environment. Second, the visual transition from the courtyard space to the interior space of the dwelling should be smooth and seamless.

Based on this requirement, traditional Chinese architectural dwellings, especially in the maintenance structure of the front eaves of the facade, often employ a combination of wooden doors, windows, partitions, and sill walls. Even when walls are present, they are typically low and made of sills. The doors and windows are intricately carved and recessed, and the interior and exterior of the dwelling share the same design elements. This maximizes the visual appeal and experiential satisfaction of both the interior and exterior spaces, creating a facade that combines functionality and aesthetics.[1] In fact, the use of such treatment in the maintenance structure of the front eaves is also a necessity for the dwelling itself. It serves purposes such as lighting, ventilation, and spatial transition. "The recessed planes created by the framework and doors and windows are positioned behind the larger eaves and platform, not only for compositional purposes but also to protect the wooden structure from direct exposure to wind and rain."[2] Undoubtedly, this treatment plays a crucial role in the transition from the courtyard space and the exterior facade to the interior space.

In terms of the layout and relationship of the group buildings in the courtyard, the four walls of the courtyard are mainly formed by the front eaves facades of the dwellings. Even the chambers on the left and right are enclosed by the front eaves facades rather than having two walls facing the courtyard. The rear eaves facades of the buildings in front of the courtyard are used as passageways and are generally treated in the same form as the front eaves facades facing the courtyard.

Chinese building facades viewed in the courtyard.

Enclosure of the Building

Front eaves enclosure

The front eaves enclosure structure of the building is of utmost importance in terms of visual appeal. In Tang Dynasty architecture, the front eaves were often constructed with brick walls,[3] with doors installed in the main chambers and windows in the secondary and corner chambers. The windows were relatively small and mostly straight lattices. From the Song Dynasty onward, the windows gradually became larger. The most prevalent was a combination of wooden doors and sill wall windows. In the Ming and Qing dynasties, the combination of wooden doors and partition walls was increasingly used. Other various forms of door and window combinations are also used.

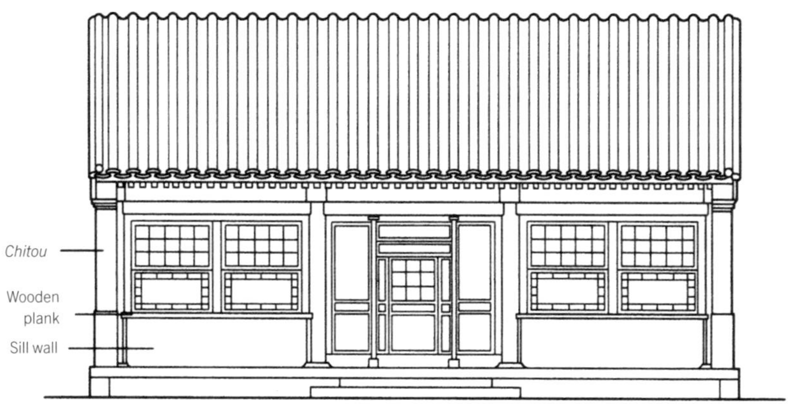

The front eave enclosure with wooden door and sill wall window combination.

TOP LEFT & TOP CENTER The front eave enclosure with wooden door and sill wall window combination.

TOP RIGHT The main hall front eave enclosure of Cuifujun Temple in Zhangzi County, Shanxi Province.

BOTTOM Front eave sill wall building approach.

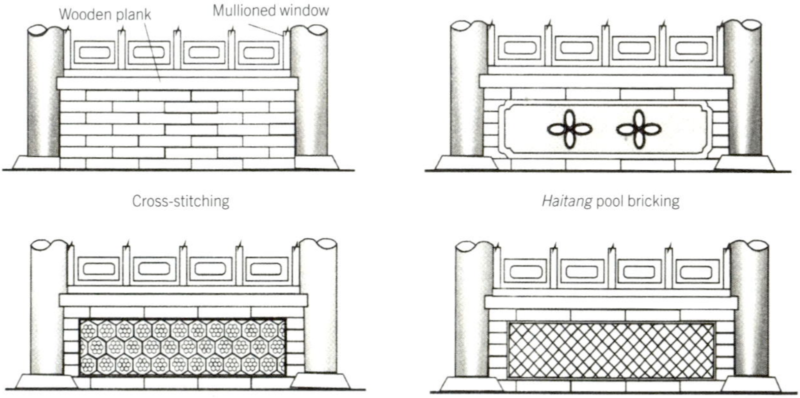

Cross-stitching

Haitang pool bricking

Glazed brick

Brick heart-shaped paneling

Wooden door and sill wall window combination enclosure. It refers to the installation of wooden doors in the main chambers, while low walls (sill walls) are constructed in the other chambers, and sill wall windows are installed on top of these walls. In the early period, the sill walls were slightly higher, while in the Ming and Qing dynasties, they were generally one-third the height of the eaves pillars. The exterior decoration of the sill walls included various techniques such as cross-stitching, *haitang* (crabapple blossom) pool bricking, glazed brick walling, and brick heart-shaped paneling.

Wooden door and partition fan combination enclosure. It refers to the installation of wooden doors in the main chamber and partition fans in the other chambers to form a front eave enclosure structure, which is commonly used in Ming and Qing dynasty-style architecture.

Another combination method is to install wooden doors in the main chamber, use partition fans in the secondary chamber, and use sill wall windows in the corner chambers and end chambers.

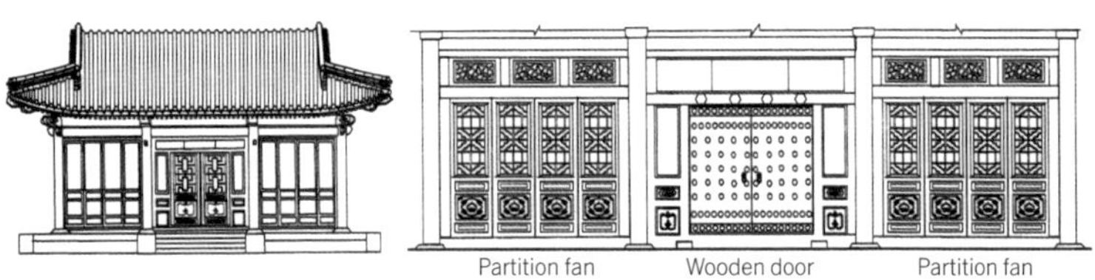

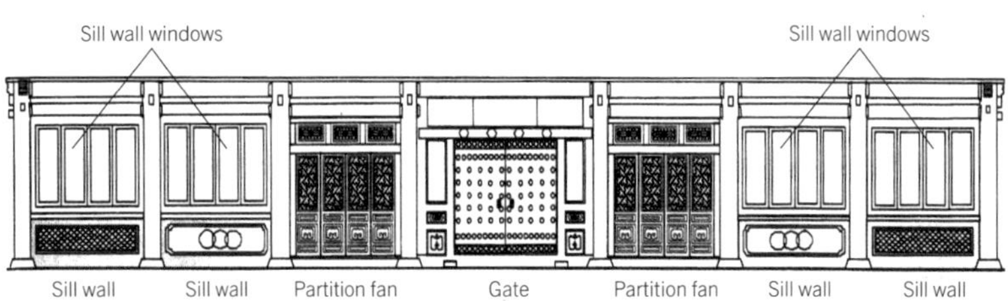

TOP Wooden door and partition fan combination for front eave enclosure.

BOTTOM The combination of wooden doors, sill wall windows, and partition fans for the front eave enclosure.

Rear eave enclosure

Rear eave enclosure refers to the rear facade of a house. If there is a passageway on the rear facade, the entire house also serves as a transition to the backyard buildings. It is designed in a similar form to the front eave enclosure, as mentioned earlier, to maintain a visual aesthetic that harmonizes with the surrounding yard walls. If it is not used as a passageway and does not have an open door, there are generally two ways to deal with it.

"Exposed eave" rear eave wall. It is constructed by laying bricks below the eave beam, with a waterproof layer on top. It is designed with a pointed and raised eave, allowing the eave beam and beam head to be exposed. The wall is divided into two parts: the upper body and the lower body (also known as "group shoulders"). The exposed eave is commonly used in hipped roofs, gable roofs, and suspended roofs. The rear eave wall can be constructed with either wrapped columns or exposed columns, and windows may or may not be installed in the rear eave wall.

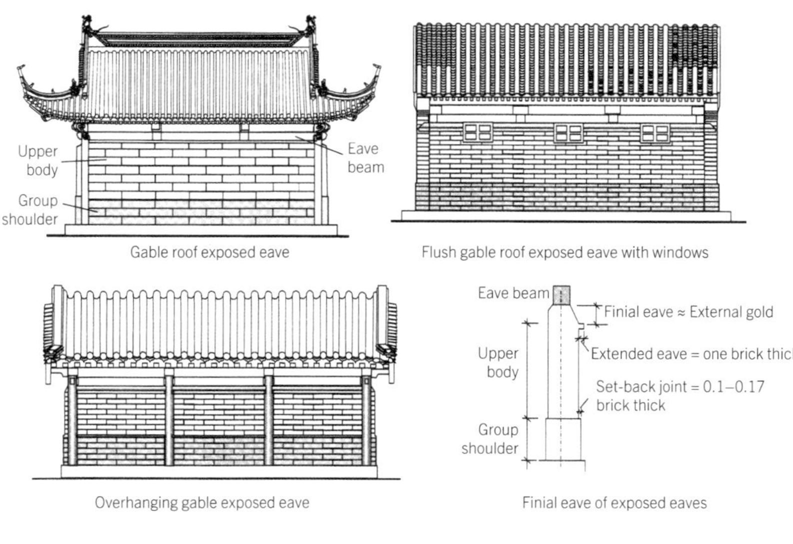

"Exposed eave" rear eave wall.

LEFT "Exposed eave" rear eave wall.

RIGHT "Exposed eave" rear eave wall of Nanchan Temple, Wutai Mountain, Shanxi Province.

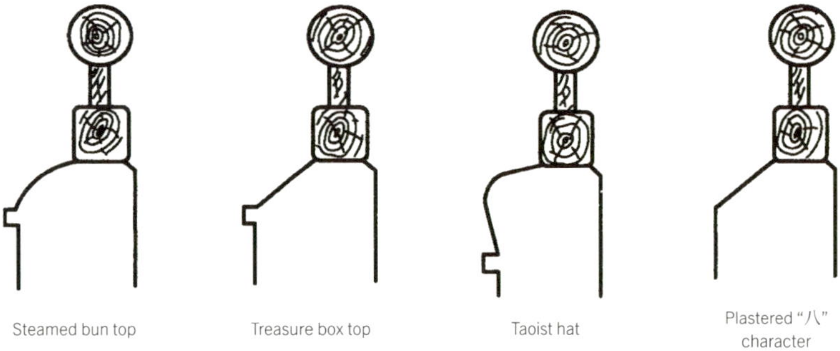

Some methods of making the eaves wall sign pointed out after the "exposed eave."

“Exposed eave” rear eave wall.

"Enclosed eave" rear eave wall. It refers to the construction of a brick wall that fills the rear eave up to the underside of the roof. The eave beam is enclosed within the brick wall, creating a solid brick wall. In addition to the "group shoulders" and the upper body, the rear eave brick wall can also have layered projecting eaves. These brick eaves can take various forms, such as straight eaves, drawer eaves, diamond-shaped eaves, chicken gizzard eaves, and ice plate eaves.

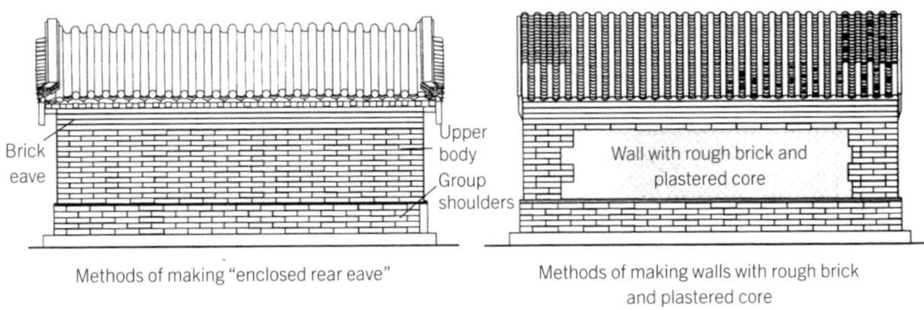

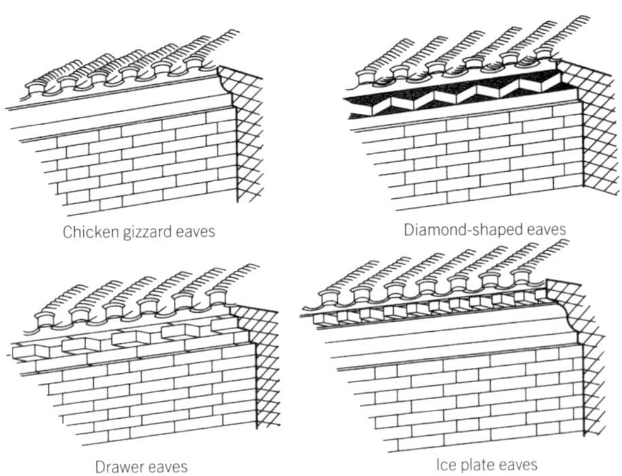

TOP Methods of making "enclosed eave" rear eave wall.

BOTTOM Various "enclosed eave" bricks.

Double mountain enclosure

Double mountain enclosure refers to the enclosure of the mountain surfaces on both sides of a house. The partition fans are used to enclose the entire mountain surface, often seen in houses with wraparound corridors, where the mountain surfaces are clearly visible. The enclosure of the mountain surfaces on both sides is often constructed in forms similar to the "exposed eaves" or "enclosed rear eaves" rear eave walls. However, when the "enclosed rear eave" form is used on mountain surfaces, it is referred to as an "enclosed mountain" enclosure.

For the "exposed eaves" of the double mountain walls, this means that the brick wall is built up to the eave beam of the mountain surface, exposing the beam to the outside. Similarly, for the "enclosed mountain" walls, the brick wall is constructed fully from bottom to top, creating a closed brick wall for the entire mountain surface.

Hip-roofed and gable-hip roofed walls are often constrained by the roof structure and, therefore, often adopt the "exposed eaves" enclosure structure.

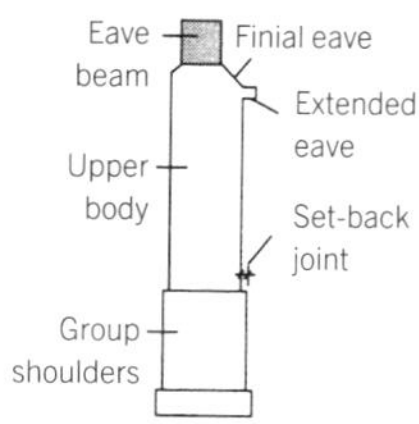

Gable section

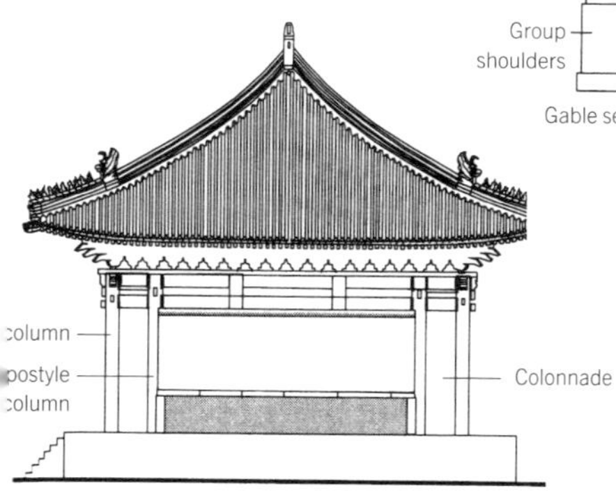

"Exposed eave" mountain walls of hip roof

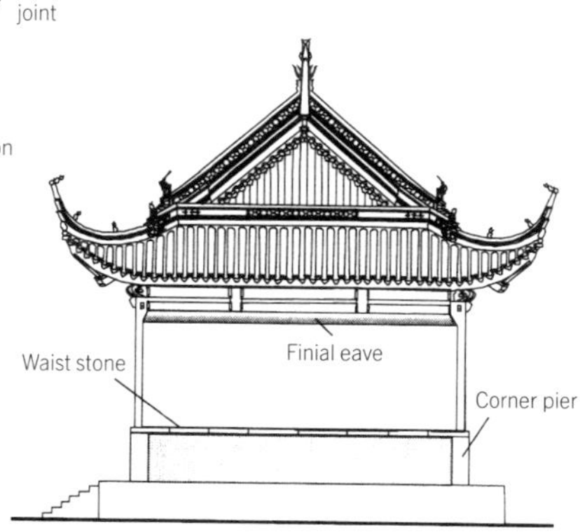

"Exposed eave" mountain walls of gable and hip roof

Mountain walls of hip roof as well as gable and hip roof buildings.

"Exposed eave" mountain walls of Kaishan Temple, Gaobeidian, Hebei Province.

The mountain walls of flush-gable-roofed buildings often adopt the "enclosed mountain" enclosure structure. In southern residential buildings, a "firewall" structure (also known as a horsehead wall) is also used. Due to the architectural form, the mountain walls of rammed earth buildings have an additional *chitou* (gable wall head) structure, which extends the width of the mountain wall beyond the position of the front and rear eaves. This is also a characteristic feature of rammed earth buildings. The *chitou* structure does not exist in the Song Dynasty architecture, but it is commonly used in Ming and Qing Dynasty buildings. The *chitou* structure consists of group shoulders, upper body, and *pantou* (capping beam). In southern residential buildings, *pantou* is also known as *duotou*. *Pantou* is the upright facade of the gable end, typically adorned with *heyedun* (lotus leaf pedestal), *banhun* (half-mixed decorations), *lukou* (furnace opening), *xiao* (owl), and *qiangyan* (eaves beam). These structures form a five- or six-layered, gracefully stretched form.

"Exposed eave" mountain walls.

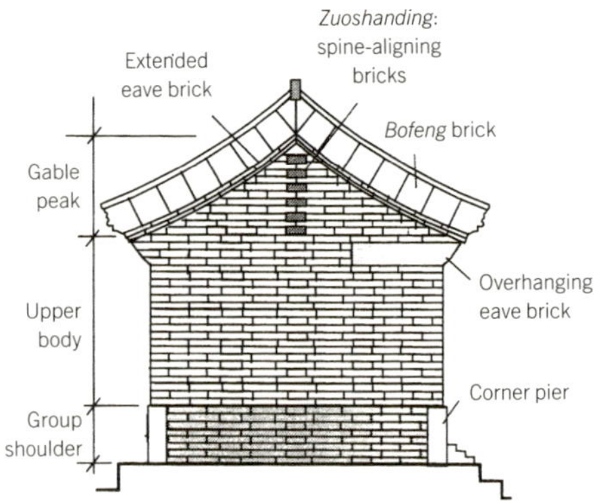

Names of various parts of gable walls

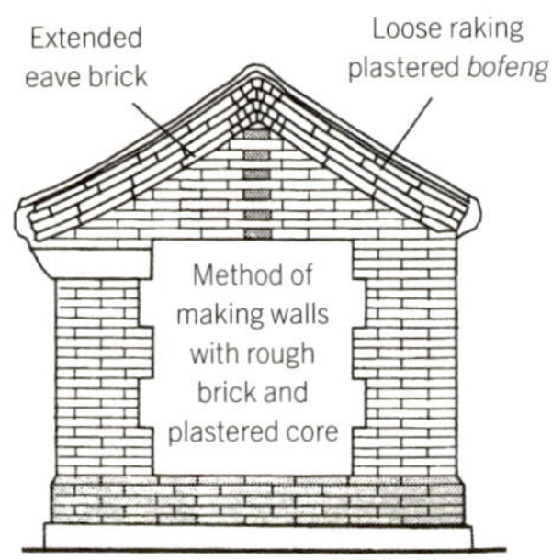

Loose raking plastered *bofeng*

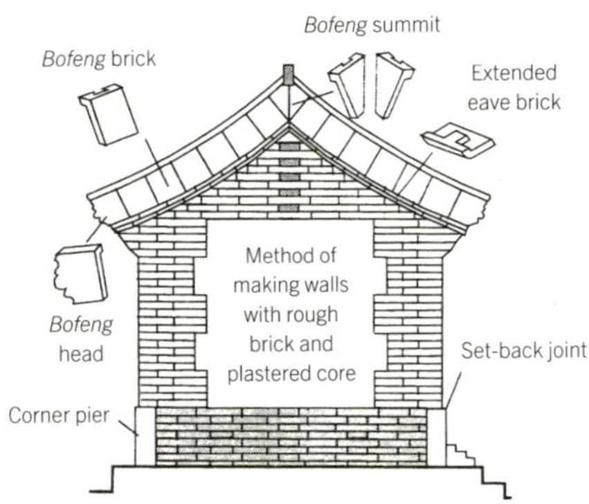

"Five in and five out" method

"Enclosed mountain" enclosure structure of mountain walls of flush-gable-roofed buildings.

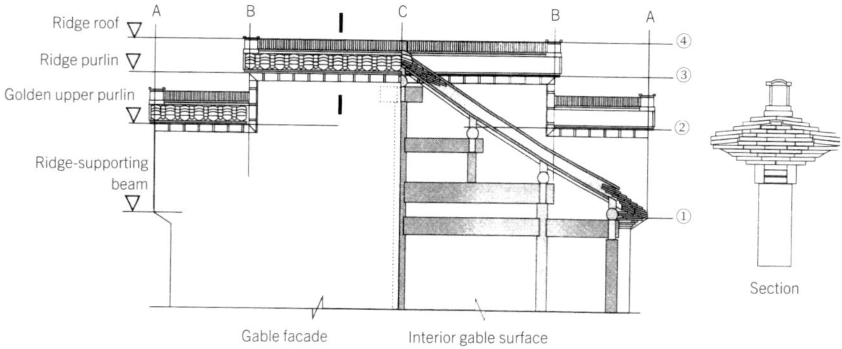

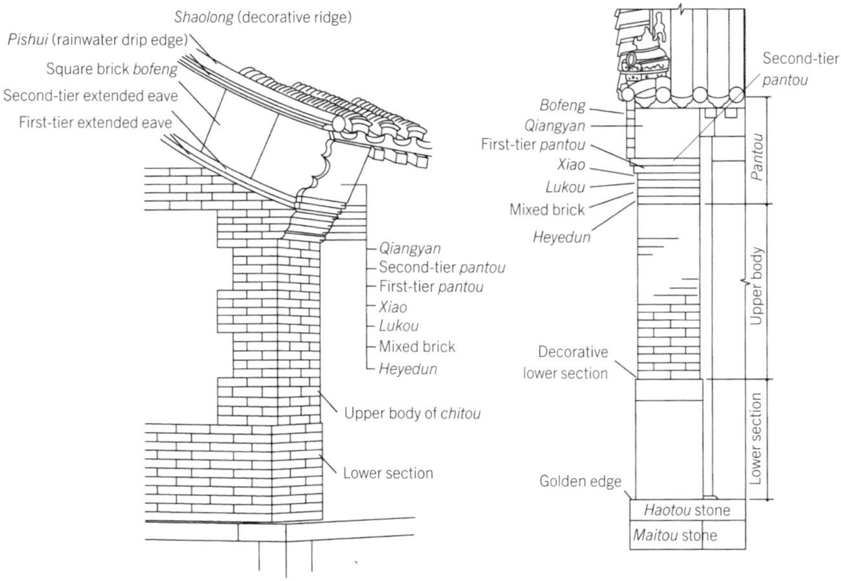

TOP Horsehead wall structure of mountain walls of flush-gable-roofed buildings.

BOTTOM *Chitou* structure of flush-gable-roofed buildings.

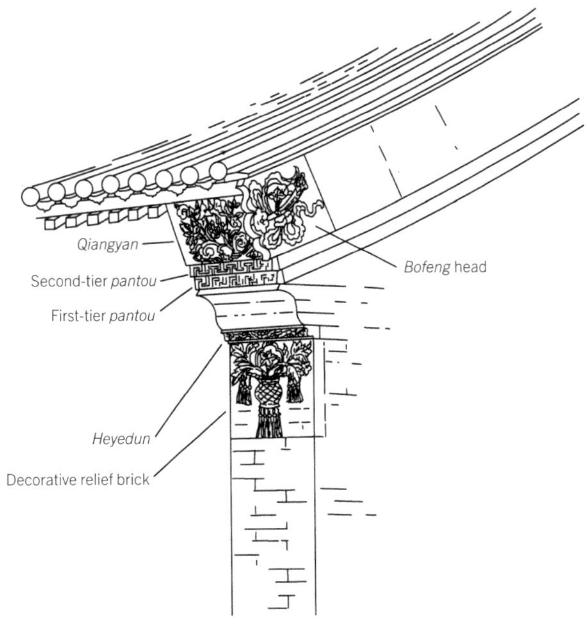

Chitou brick carving.

The mountain walls of overhanging gable-roofed buildings often adopt the "exposed eave" or "enclosed mountain" approach. Depending on the region and climate where the house is located, in the northern areas, it is common to adopt an "enclosed mountain" or "exposed eave" method. In southern residential architecture, there are two common practices. One is to brick the upper body wall below the bottom beam, exposing the upper beams and short columns, and covering the gaps between the beams with wooden boards, known as the wind-blocking board method. The other is to brick the upper body wall below the bottom beam and then brick the gaps between the beams and short columns inside, exposing the ends of the beams and short columns, which is known as the "*wuhua* mountain wall" (stepped gable wall) method.

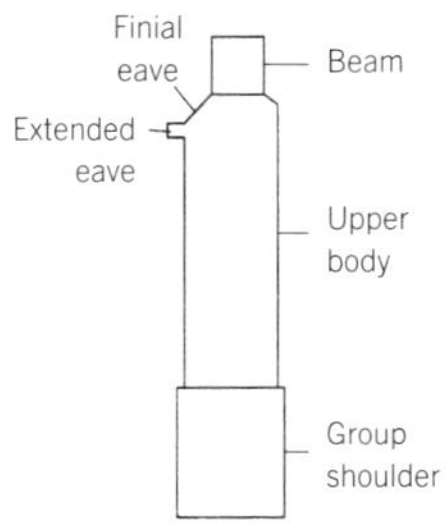

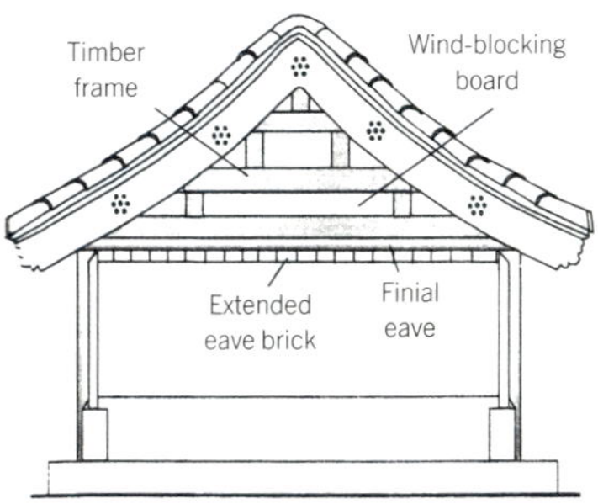

Wind-blocking mountain wall

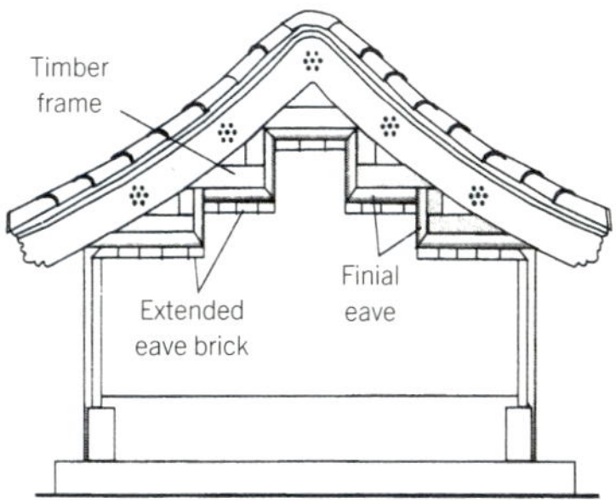

Wuhua mountain wall

Mountain walls of overhanging gable-roofed buildings.

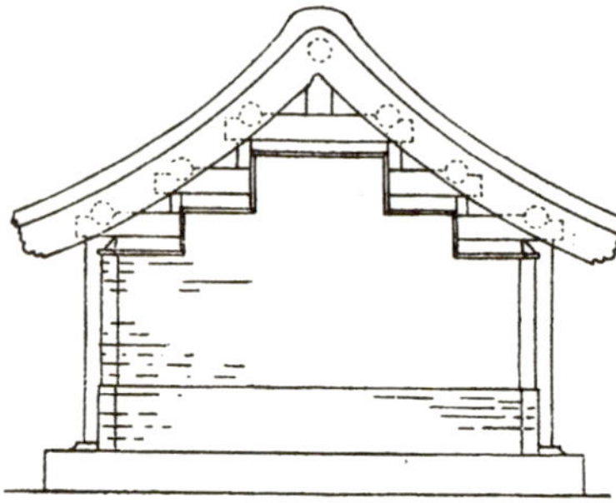

Wuhua mountain wall.

Bofeng board and *xuanyu* of Zhenru Temple, Shanghai.

Mountain Surface Decoration

The triangular space between the two slopes of the upper part of the mountain walls on both sides of ancient buildings is called the *shanmian* or *liangmian* (mountain surface). The mountain surface can be either enclosed or open, and it is often adorned with woodcarvings and brick carvings, also known as *shanhua* (mountain flowers), representing a floral motif in architectural decoration.

The decorative elements of *bofeng* board and *shanhua* board. *Bofeng* board is a zigzag-shaped protective board that covers the protruding ends of the rafters on the mountain surface of overhanging gable roof or gable and hip roof buildings. It was called *bo* (fight) *feng* (wind) in the Song Dynasty, which means to fight against wind and rain. It is also referred to as *bo feng* (gap) board (gable eave board) by later generations. For flush gable roof buildings, *bofeng* bricks are used on the mountain surface. *Bofeng* bricks are made by assembling large square bricks, and the bottom brick is called *bofeng* head, which can be plain or decorated. The construction method of *bofeng* board is described in the Song Dynasty *Yingzao Fashi*, "To make *bofeng* board, install it outside the tenon/nail heads between the two sides of the roof, with a width of two to three materials and a thickness of three to four parts, and a length according to the frame. The upper part of the framework should have inclined brackets on both sides, with a length of two feet five inches to three feet, and the lower part should be level with the rafters and tile heads." According to the Qing Dynasty *Gongcheng Zuofa Zeli*, the *bofeng* board should be made to match the length of each rafter, with a width determined by six parts of the rafter's diameter, and a thickness equal to that of the *shanhua* board. When the *bofeng* board was installed in the Qing Dynasty, a groove was carved on the back of the board to insert the rafter head, ensuring that the *bofeng* board would fit tightly against the *shanhua* board.

The *shanhua* board is a triangular wind and rain shield on the mountain surface of a gable and hip roof building. The top two sloping sides of the *shanhua* board are level with the surface of the rafters. A groove is carved on the *shanhua* board to allow the rafters to pass through. *Shanhua* boards come in two styles: plain painted or adorned with gold leaf and painted decorations. The painted motifs include *wanhua* (round floral patterns) and *shoudai* (silk ribbons tied to an official seal), symbolizing wealth and auspiciousness. In traditional residential architecture, brick carvings are often used to depict auspicious floral motifs on *shanhua* boards.

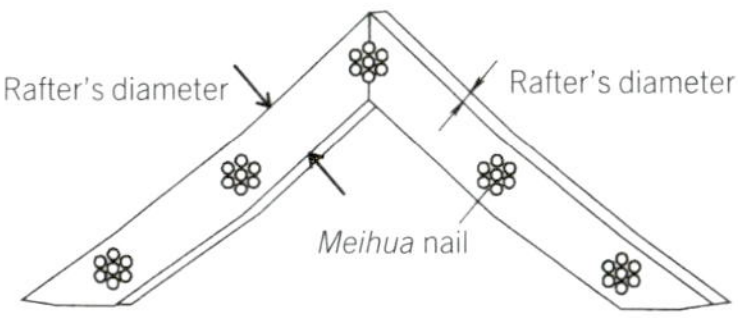

Bofeng board and *meihua* nail

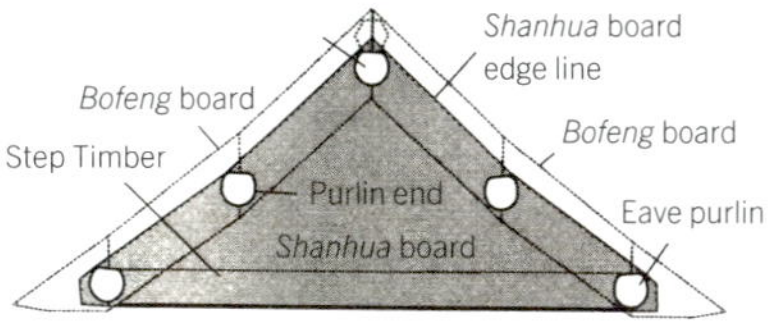

Shanhua board

TOP *Bofeng* board, *shanhua* board, and *meihua* nail.

BOTTOM *Bofeng* board.

***Xuanyu*, *recao*, and *meihua* nail.** Below the *bofeng* board, there are *xuanyu* (also known as *chuiyu* in the Song Dynasty) and *recao*. The *xuanyu* is located under the sharp corner of the mountain surface of the *bofeng* board, while the *recao* is positioned under the skinning wood of the *bofeng* board. Both the *xuanyu* and *recao* can be decorated with petal patterns, cloud patterns, and so on. The tradition of *xuanyu* is derived from the story of the "Prefecture *Xuanyu*" (the prefecture who hung the fish). In *Book of the Later Han · Biography of Yangxu*, it is recorded, "During his tenure as the Prefecture of Nanyang, Yang Xu exhibited integrity and self-discipline. When his assistant presented a live fish as a gift, Yang Xu hung it up in the courtyard as a sign of refusal, showcasing his incorruptibility." Hence, he earned the nickname the "Prefecture *Xuanyu*." In addition, both fish and *recao* (a type of aquatic plant) thrive in water. So besides the symbolic meaning of integrity, some believe that it also represents a love of water and fire prevention, similar to the aquatic plant decorations in the caisson ceiling. In the architecture of the Qing Dynasty, there was a practice of using *meihua* nails (plum blossom nails) as decorative elements in the position opposite to the decorative wooden brackets on the outer facade of the *bofeng* ban. Plum blossom nails are wooden blocks shaped like plum blossoms, and their size varies according to the architectural design and the size of the *bofeng* board. A set of seven plum blossom nails is usually used for decoration.

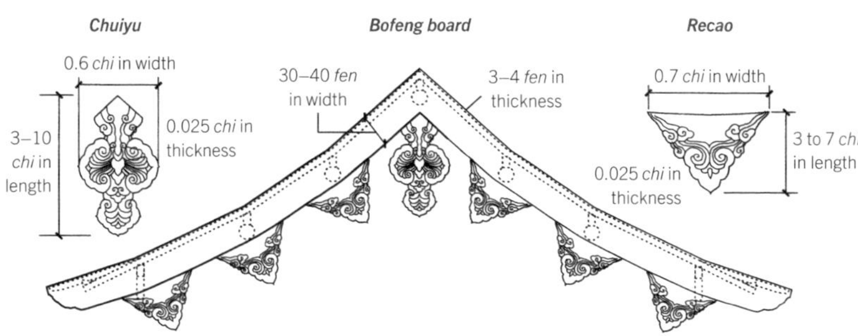

Xuanyu, recao.

TOP LEFT, TOP RIGHT, MIDDLE LEFT, MIDDLE RIGHT *Xuanyu, Recao.*

BOTTOM LEFT *Xuanyu* and *recao* in *Yingzao Fashi.*

Walls

Walls are barriers and enclosing structures both inside and outside buildings, and they vary widely in terms of materials, scope, and form. Based on the materials used, there are various types of walls, including rammed earth walls, adobe walls, brick walls, stone walls, wooden walls, and walls made of woven strips of mud. There are also walls that combine multiple materials, such as walls with a stone base and a top made of earth or brick, walls with a brick base and a top made of earth, or walls with a hollow core and a brick facing with fill soil in between.

Rammed earth walls are one of the oldest forms of walls in China.[4] In rammed earth, wooden boards are used as molds to compact layers of clay or a mixture of clay (often mixed with lime or sand and gravel). This technique, also known as *banzhu* (board construction), has been found in architectural sites from the Shang, Zhou, Han, and Tang dynasties. The *Yingzao Fashi* provides specific guidelines for wall construction, including the requirement for a tapering thickness. The height of a rammed earth wall is usually twice its base thickness, and the top thickness is about 1/5 to 1/4 of the wall height. However, in actual house construction, the height of rammed earth walls is usually not very high, so the tapering thickness is reduced correspondingly.

Brick is a widely used material for walls in ancient architecture. The use of bricks has been discovered in archaeological sites dating back to the Warring States period. The brick pagoda of Songyue Temple of Northern Wei Dynasty, Henan Province, indicates that brick production and use had reached a high level during that time. The *Yingzao Fashi* provides specific descriptions of brick production and bricklaying techniques. There are various methods of bricklaying, such as horizontal half brick, diagonal flat diagonal brick, horizontal side brick, horizontal diagonal brick, horizontal vertical brick, and vertical diagonal brick. During the Ming and Qing dynasties, the most common brick patterns for walls were

Rammed earth walls.

three horizontal and one diagonal, two horizontal and one diagonal, or one horizontal and one diagonal. Brick walls can also be hollow, such as hollow brick walls or cavity walls. Hollow brick walls are constructed by stacking bricks to form a hollow box shape, which is then filled with a mixture of soil. These walls are often non-load-bearing and can be constructed using various techniques, such as horse trough blocks, stacked box blocks, or tall and short blocks.

In traditional timber houses in the southern regions of China, wood is commonly used for both exterior and interior walls. Woven bamboo strips with mud walls are also used for both exterior and interior walls. Stone walls, except in some mountainous areas or for small buildings and tombs, are not as commonly used in residential construction as brick walls and rammed earth walls. However, there are some houses that have walls made of a mixture of stone, rammed earth, and brick.

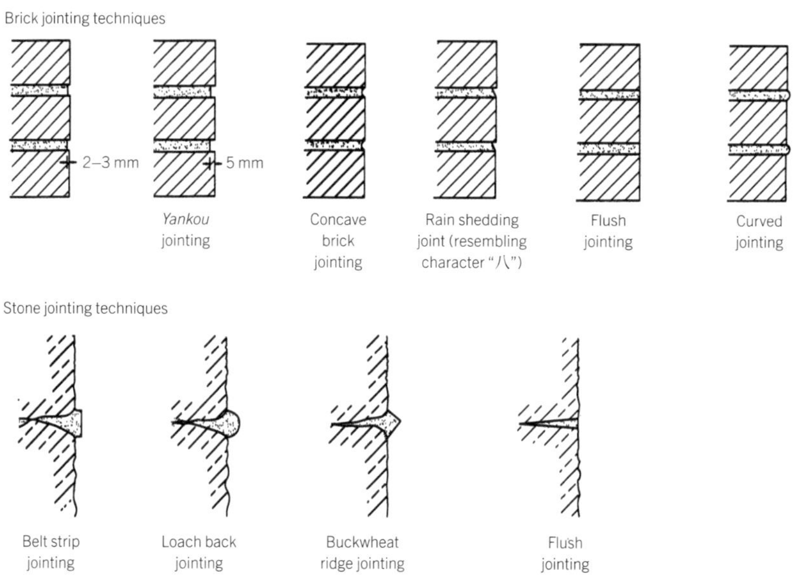

Brick jointing and stone jointing techniques.

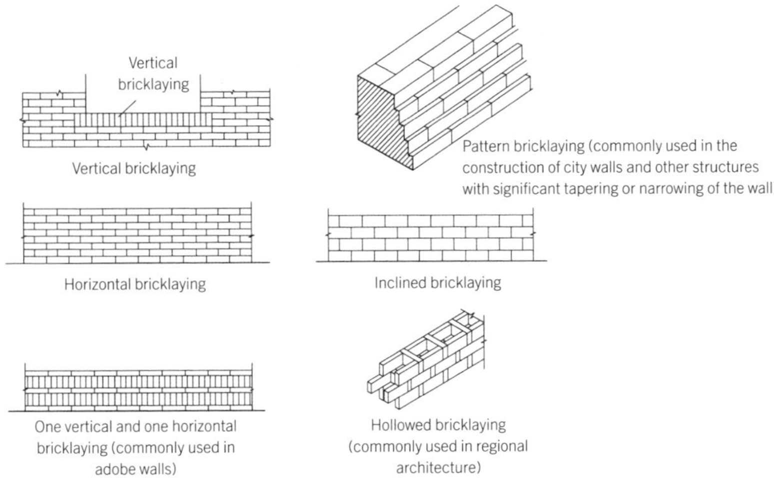

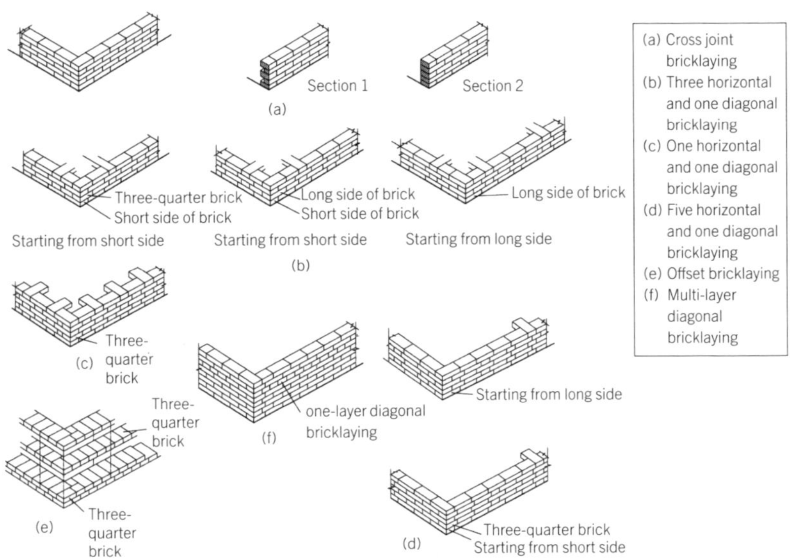

TOP Bricklaying.

BOTTOM Horizontal bricklaying.

NOTES

1. Mr. Li Yunhe believes that due to the adoption of a frame structure, ancient Chinese architecture has the earliest existence of a curtain wall structure that consists entirely of doors and windows.
2. Li Yunhe, *Principles of Classical Chinese Architectural Design* (Tianjin: Tianjin University Press, 2005), 181.
3. Some parts of the architecture use rammed earth walls and adobe brick walls.
4. Pan Guxi, *The History of Chinese Architecture* (Beijing: China Architecture & Building Press, 2015), 282.

CHAPTER 6

ROOF

Roof Forms

The great roof is the most prominent element in the overall composition of ancient Chinese architecture, boasting the largest scale and captivating the visual senses. Unlike Western architecture, which tends to hide roofs behind walls, ancient Chinese architecture showcases the great roof as a defining element. Chinese characters, such as "palace," "chamber," "residence," "home," "wing room," "corridor," "courtyard," "mansion," "shop," and "storehouse," all related to dwellings, are invariably adorned with a roof, reflecting the profound appreciation the ancient people held for the shelter it provided.

Among the three components of a house, namely the platform, the facade, and the roof, it is the roof that garners the utmost attention in terms of its external appearance. The treatment of roofs also accentuates the presence of a structure within an architectural ensemble. The myriad roofs in ancient Chinese architectural clusters coalesce to create a distinctive panorama in the realm of Chinese antiquity.

Ancient Chinese architecture has developed a variety of roof forms over the centuries. Some of the typical roof forms include the *wudian* (hip roof), *xieshan* (gable and hip roof), *xuanshan* (overhanging gable roof), *yingshan* (flush gable roof), *juanpeng* (round ridge roof), and *cuanjian* (pyramidal roof). In the *Brief Record of the Palace · Explanation of the Principles of Roof Construction*, it states, "The imperial coffin is transported on a dragon carriage. Upon the coffin's casket, there is a canopy known as *guo,* upon which a tall structure called *zouju* is placed. The wooden structure surrounds it like a house with four sides extending and meticulously covered. In *The Rites of Zhou · Record of Crafts and Artisans*, it is recorded, "殷人重屋, 堂脩七寻, 堂崇三尺, 四阿重屋" (For *chongwu* in the Yin Dynasty, the depth of *tang* was seven *xun*, the height three feet, hip-roofed and double-eaved). According to Zheng Xuan's annotation, the "四阿" (sì ē) is similar to the contemporary concept of "四注" (sì zhù), a four-cornered hip-roofed structure or house. Jia Gongyan's commentary adds that these "四阿" (sì ē) are also known as "四霤" (sì liù). The terms "四阿" (sì ē), "四注" (sì zhù), and "四霤" (sì liù) mentioned here all refer to the four-sided sloping roofs, also known as *wudian* or *wuji* roof. These descriptions are among the earliest known references to roof forms in literature. The four-sided sloping roof and the inverted-V roof or hipped roof are the origins of various other roof styles. Throughout the development of ancient architecture, numerous roof forms have evolved based on the foundation laid by the four-sided sloping roof and hipped roof.

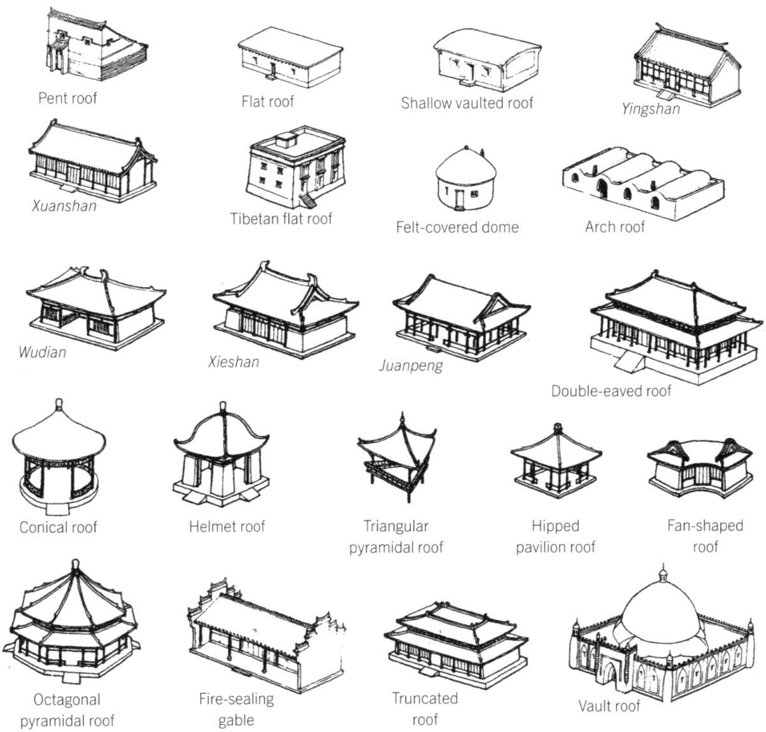

Roof forms of ancient Chinese architecture.

The *wudian* roof, also known as the four-eave roof or five-ridge roof in the Song Dynasty, is the highest form of roof in ancient architecture. It is commonly used for the main halls in imperial palaces and temples. It can have a single eave or double eaves, with the latter being grander. The single-eave *wudian* roof has one main ridge and four vertical ridges at the corners, making a total of five ridges. Therefore, the single-eave *wudian* roof is also known as the *wujidian* (five-ridge hall). The earliest written record of the *wudian* roof can be traced back to the Shang Dynasty. The architectural form of the *wudian* roof can be seen in Zhou Dynasty bronze vessels, Han Dynasty stone carvings, and the stone caves of the Southern and Northern Dynasties. The earliest surviving wooden structure with a *wudian* roof is the main hall of the Foguang Temple. During the Qing Dynasty, there was a special technique called *tuishan* used in the construction of the *wudian* roof. It involved extending the ends of the central ridge, pushing the two peaks outward. This adjustment was made to accommodate narrower spacing between the roof supports and prevent the main ridge from being too short, which could result in a less pronounced curve for the vertical ridges and a straighter appearance. By extending the main ridge, the curve of the secondary ridges became more graceful and elongated. By extending the central ridge, the

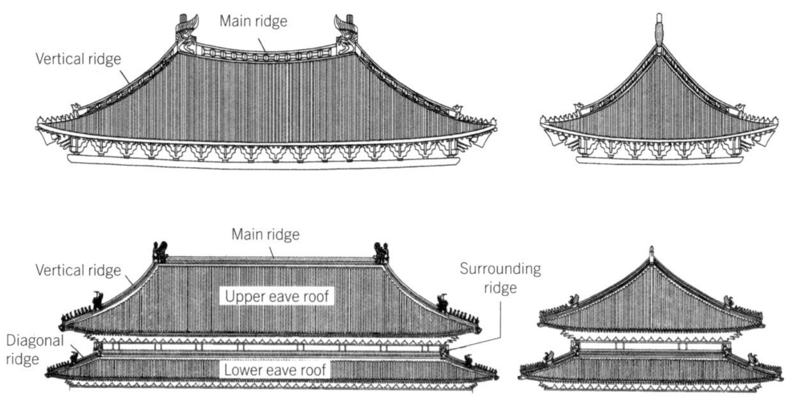

TOP Single-eaved *wudian* roof.

BOTTOM Double-eaved *wudian* roof.

rafters were extended beyond the beam structure, and the ends of the rafters were supported by suspended pillars. Under the suspended pillar, a *taiping* beam was used, resting on front and back gold beams.

The *xieshan* roof, also known as the nine-ridge hall in the Song Dynasty, is a roof form that ranks second in the hierarchy to the *wudian* roof. It is composed of a main ridge, four vertical ridges, and four diagonal ridges, hence the name "nine-ridge hall." The earliest surviving wooden example of the x*ieshan* roof is the Main Hall of the Nanchan Temple in Wutai Mountain. According to Professor Liu Zhiping, the *xieshan* roof was originally a variation of the inverted-V roof. Later, an additional surrounding corridor was added around the inverted-V roof (hipped roof), resulting in a gable and hip roof. This viewpoint is supported by evidence found in the stone carvings in the *que* of the Han Dynasty and the Main Hall of the Dongchang Temple in Huoxian County, Shanxi Province.[1] The *Examples of Qing-Style Construction* explain that the *xieshan* roof is a combination of the *xuanshan* and the *wudian*. The upper half of the vertical ridge, from the main ridge to the junction with the secondary ridges, is structurally identical to the *xuanshan*. The lower half is identical to the *wudian* roof, from *bofeng* to *xianren* (celestial being), with the same arrangement of animals in the front and

Single-eaved *wudian* roof.

back. The lower half, from *bofeng* to *taoshou* (a set of animal images), is called the diagonal ridge, which forms a 45-degree angle with the secondary ridges in the plane. The surface of the *xieshan* roof is adorned with various decorative elements, such as the *bofeng* board and *xuanyu*. The mountain surface is often provided with protective seam strips, and it can be embellished with carved designs, painted decorations, or even windows, resulting in a wide range of variations in appearance.[2]

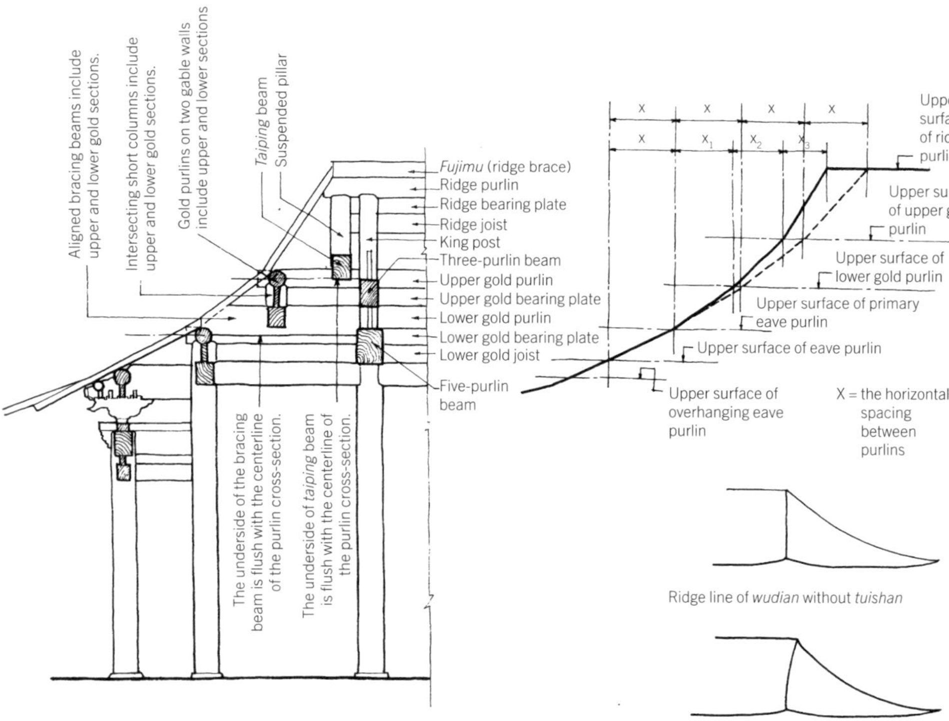

Longitudinal section of the structure frame of *wudian*'s end bay

Ridge line of *wudian* without *tuishan*

Ridge line of *wudian* with *tuishan*

Tuishan technique in the construction of *wudian*

Tuishan technique used in the construction of *wudian* roofs during the Qing Dynasty.

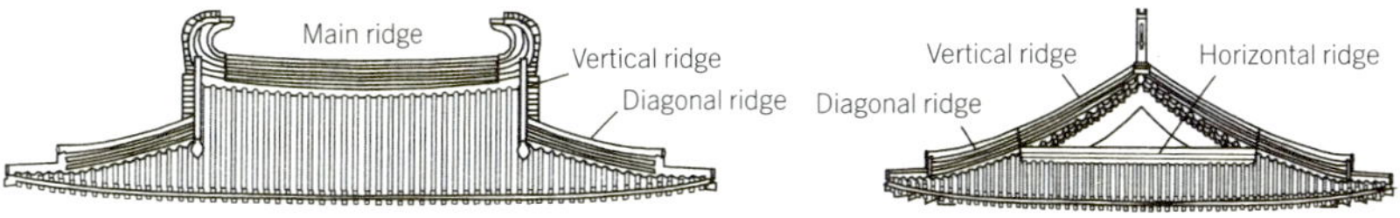

The front and side views of the single-eaved *xieshan* roof in the Song Dynasty

Single-eaved *xieshan* roof.

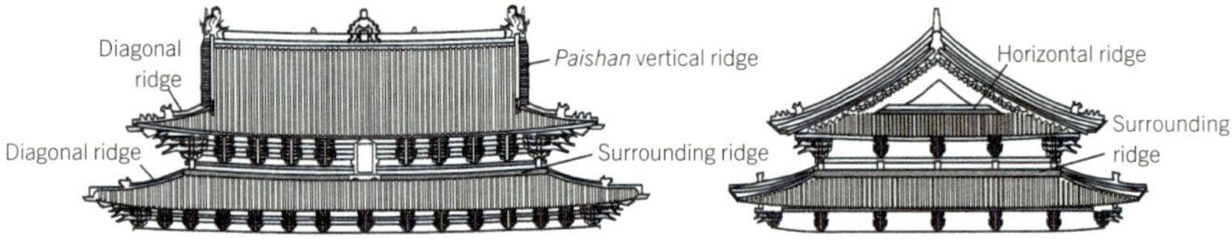

The front and side views of a double-eaved *xieshan* roof in the Song Dynasty

Double-eaved *xieshan* roof.

Xieshan roof with rolled ridge.

The *xuanshan* roof is a type of roof with two-sloping sides. It is a more common roof form, as reflected in murals, stone carvings, and Han Dynasty portrait stones. It is generally not used for important buildings and has lower specifications compared to *wudian* and *xieshan*. It is more commonly used in folk residences. The eaves of the *xuanshan* extend beyond the gable walls, and it has a central ridge and secondary ridges. The *juanpeng* roof has no central ridge. The gable walls of the building typically show the wooden framework of columns, beams, and rafters. Some may have decorated stepped gable walls, with *bofeng* board to protect the exposed eaves rafters.

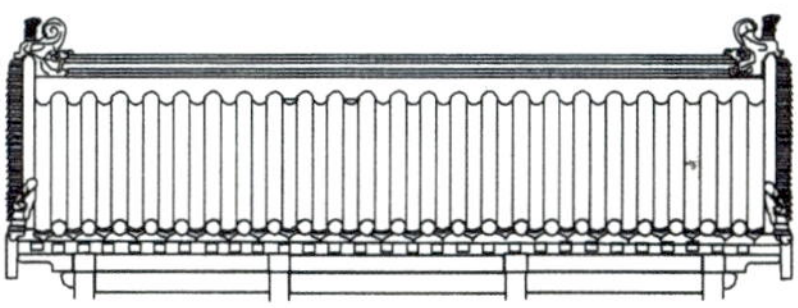

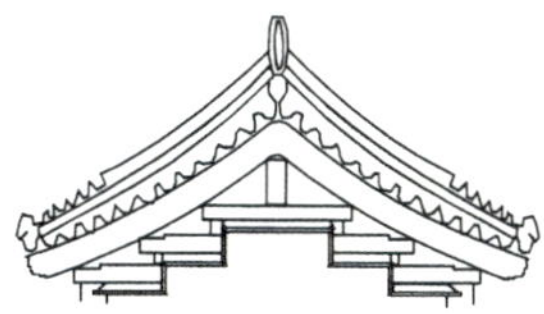

Xuanshan roof.

Xuanshan roof.

The *yingshan* roof is another type of double-sloped roof, but unlike the *xuanshan* roof, the eaves of the *yingshan* roof do not project outward. The center columns, eave columns, *paishan*, and lintels of the two slopes are all built inside the gable walls and are not exposed. Under the gable walls, the *paishan goudi* (the eagle's beak and drip groove on the upper and outer edge of the eaves) is designed and created, using *bofeng* bricks, and the gable walls are made into a *chitou*. The use of the *yingshan* roof can be traced back to the Song Dynasty, and the construction method of the gable walls is related to the extensive use of bricks. It was widely used in residential buildings during the Ming and Qing dynasties.

Yingshan roof.

The *cuanjian* roof, also known as *doujian* in the Song Dynasty, was first seen in the stone carvings of pagodas in the Northern Wei grottoes. The *cuanjian* roof has a central point where several secondary ridges converge at the top, and it is crowned with a precious top, which can be square, round, hexagonal, octagonal, and so on. The roof truss structure varies according to the plan form, with single eaves being more common, although multiple eaves can also be found on pagodas.

In fact, there are other types of roofs in addition to those mentioned. In the Shaanxi region, there is a single-sloped roof commonly seen in traditional residential buildings. In arid regions, flat roofs are often used due to the lack of rainfall. There is also the *yinding* roof, a flat roof built on the upper part of the *wudian* roof. Additionally, there is the *shiziji* (cross-ridge) roof, forming a cross-shaped ridge at the top.

Cuanjian roof.

Cuanjian roof.

Roofing Materials

In ancient Chinese architecture, various materials were used for roofing, including thatch, clay, slate, clay tiles, and glazed tiles. Official buildings often used glazed tiles and clay tiles, while residential buildings commonly used clay tiles, thatch, clay, and slate. Slate was primarily used in mountainous areas with abundant stone resources. Thatch and clay were used throughout different dynasties, but mainly by the common people and lower income classes. Clay tiles were later used more frequently in residential buildings.

Glazed tiles roofing. The use of glazed tiles on roofs is a common practice in Chinese architecture. Glazed tiles are ceramic tiles with a glazed surface, which not only enhances the aesthetic appeal but also improves water resistance and drainage. They are typically used in more prestigious buildings such as palaces and temples. During the Qing Dynasty, there were strict regulations on the use of glazed tiles. Princes and nobles used green glazed tiles with green borders, while the imperial palace and temples used yellow glazed tiles with yellow borders. Royal gardens could use a combination of different colored glazed tiles. The use of yellow glaze can be traced back to the Han Dynasty, and the use of glazed tiles was discovered during the Southern and Northern Dynasties. Glazed tiles became more prevalent from the Song Dynasty onwards, and by the Tang Dynasty, they were available in various colors, no longer limited to yellow and green. The commonly used types of glazed roof tiles include cylindrical and flat tiles, *chiwei** tiles (later changed to *chiwen* or *shouwen* tiles), hanging beast tiles, corner beast tiles, and celestial being tiles.[3]

The glazed roof tiles are composed of various tiles, including cylindrical tiles, ridge tiles, flat tiles, eave tiles, drip tiles, as well as star tiles, nail caps, and tile nails.[4] Bamboo cylindrical tiles and bamboo flat tiles with a larger front end and a smaller back end are also used in some circular or conical roofs, and the tile ridges are arranged in a radiating pattern. In addition to conventional practices, there are other techniques, such as glazed tile edging and glazed tile mosaic. Glazed tile edging involves using glazed tiles for the eaves and ridges, while using cut tiles or clay tiles for the roof surface. Alternatively, one color of glazed tiles

* A decorative sculpture installed at the ends of the roofs, possessing the shape of a bird's wing.

can be used for the eaves and ridges, while another color is used for the roof surface. Prior to the Song and Yuan dynasties, glazed tiles were only used on the ridges and not on the eaves for edging. Glazed tile mosaic, also known as "jujin," is a technique where two or more colors of glazed tiles are used to create patterns on the roof.

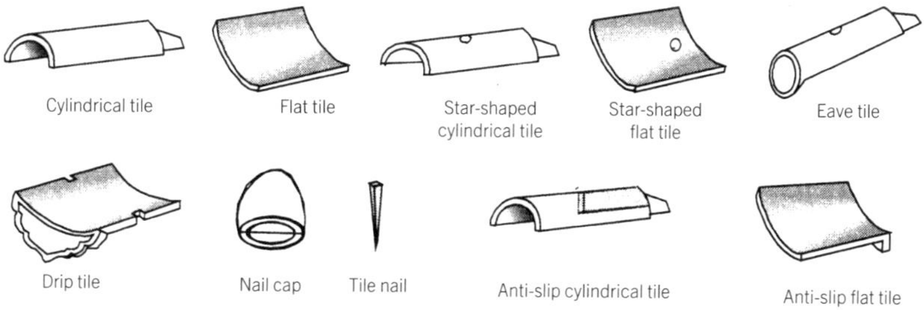

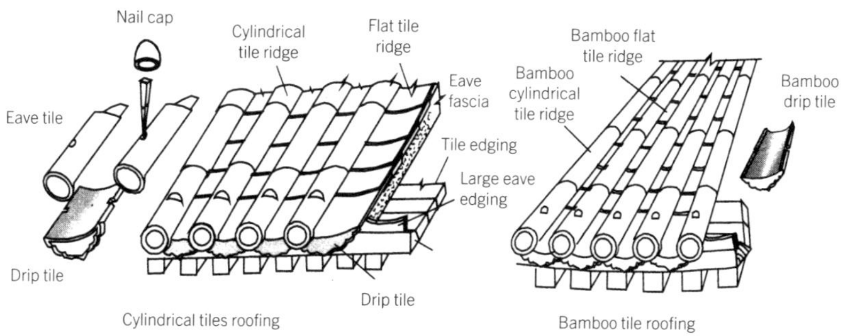

TOP Tiles of glazed roofing.

BOTTOM Glazed tiles roofing.

Glazed tiles roofing.

Clay tile roofing. Clay tile is a dark gray colored tile. It can be used for various types of roofing, such as cylindrical tile roofing, combined tile roofing, inclined tile roofing, and dry rolled tile roofing. Tube tile roofing uses flat tiles as the tile base and tube tiles as the tile cover. Combined tile roofing also uses flat tiles as the cover tiles, with the base and cover tiles arranged in alternating positions to form a tile ridge. In the northern regions, this type of tile ridge is called "*yinyang* tile," while in the southern regions, it may be called "butterfly tile" or "small green tile." The installation of the tiles can be done either by using mortar to lay them or by directly laying the cover tiles between the base tile ridge without using mortar. In a sloped tile roof, a sloped ridge is created between two base tiles using mortar, similar to tube tiles, without using actual tube tiles. Inclined tile roofs do not have complex primary and secondary ridges and are commonly used for lower-income housing. Dry rolled tile roofing does not have cover tiles, and the tiles are interlocked and pressed together without the use of mortar in between the tile ridges. Dry rolled tile roofing is cost-effective, lightweight, and offers good waterproofing, making it a distinctive roofing tile option.

In the eaves section of glazed and clay tiles, decorative edge tiles and drip tiles are used. The decorative edge tiles are called *huatou banwa* in the Song Dynasty. When flat tiles are used as a roofing tile, the end of the flat tile at the eaves is turned up in a fan shape to prevent rainwater from flowing back along the lower part. It is decorated with patterns. The Song Dynasty called the drip tiles *chujian huatou banwa*. They are placed at the bottom of the tile ridge, along the eaves of the roof, and have sharp leaf-shaped (often auspicious shapes in the Qing Dynasty) dangling tongue tiles, serving to channel rainwater and prevent water from dripping directly down.

Alternative material roofing. In addition to the aforementioned materials, there are other materials used for roofing in ancient architecture, depending on the region, climate, and other conditions. These include thatched roofs, slate roofs, and rammed earth roofs. Thatched roofs, also known as straw roofs, are covered with materials such as thatch, wheat straw, or rice straw. They have a long history of use and are commonly seen in traditional buildings. Slate roofs are made by laying thin stone slabs as the roof covering. This type of roof provides durability and a unique aesthetic appeal. Rammed earth roofs are commonly found in arid regions. They are constructed by compacting various types of soil onto a platform roof. For example, the use of Aga soil in Tibet is particularly distinctive. These alternative roofing materials were selected based on their suitability for specific regions and their ability to withstand local weather conditions.

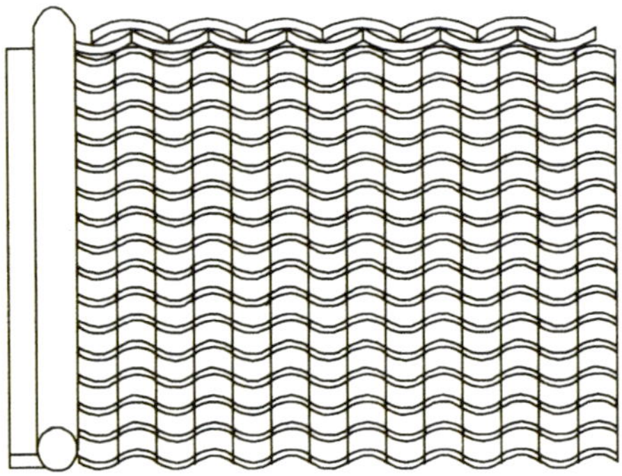

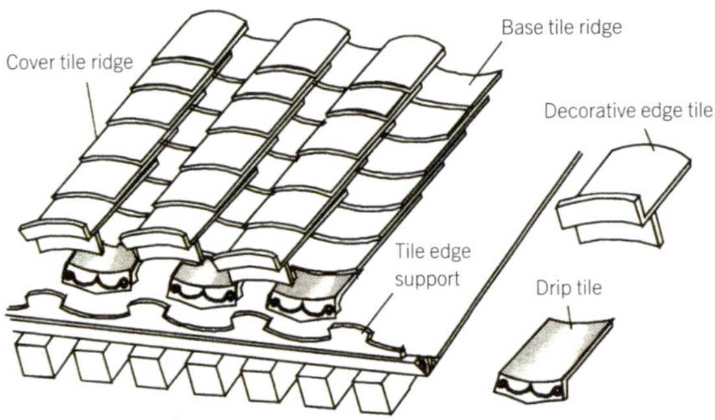

TOP Combined tile roofing.

BOTTOM Clay tile.

Roof Structure Characteristics

The special construction of the roofs in ancient Chinese architecture, as well as the formation of the curved lines in sloping roofs, is not only one of the most prominent features of ancient Chinese architecture but has also been a subject of interest for both Chinese and foreign scholars. The causes and functions of these features have been studied and interpreted by scholars from different perspectives, leading to various understandings and judgments. However, it is difficult to reach a unified and definitive conclusion.

Lin Huiyin once said, "The roof curves, which have always been regarded as extremely unique and mysterious, do not go beyond the structural and natural elements, and they are very successful in terms of both aesthetics and functionality. The entire curve of the roof slope transforms the originally dull and clumsy roof into a beautiful crown that adorns the entire building, with its upper part rising proudly and the eaves gently spreading out like wings."[5]

Le Jiazao said, "The curved lines on Chinese great roofs were not intentionally designed. When we observe older thatched roofs or old tiled roofs, we often see a downward curve in the middle of the roof surface. This is where the curved lines originate. Over time, the degree of curvature increases. It can be understood that the curved lines on the roofs initially resulted from the weaknesses in technology and materials rather than for aesthetic purposes. Later on, people embraced and utilized these curves for their beauty, and the raised edges and corners were natural consequences."[6] Le believes that the formation of curved lines on ancient Chinese roofs was the result of natural sagging and deterioration and the subsequent utilization of these curves for aesthetic purposes. This understanding clearly fails to recognize the precise wooden construction techniques used in creating the curved roof structures.

Liu Zhili believes, "The reason why Chinese roofs have concave curves is mainly due to the presence of multiple columns. The different heights of these columns cannot be drawn as a straight line, so it is preferable to gradually raise them to create concave curves, in order to avoid unevenness on the roof surface. Over time, we have come to appreciate the beauty of these concave curves."[7]

In *The History of Chinese Architecture*, Japanese scholar Ito Chuta believes that the curved lines on Chinese great roofs were mainly influenced by the shape of nomadic tents and the merging of primary and secondary structures. On the other hand, Mr. Joseph Needham believes, "Regardless of our thoughts on tents, the upwardly curved eaves in China clearly have practical effects in capturing sunlight in winter and reducing heat in summer. It can lower the height of the roof while maintaining a steep slope and provide a wide span at the eaves, thereby reducing lateral wind pressure. The fact that the columns are simply placed on stone foundations rather than set into the ground is critical in preventing their movement. Another practical effect of the downward curved roof is that it can direct rain and snow away from the foundation and into the courtyard."[8]

The opinions mentioned above have proposed possible causes for the structural forms of roofs in ancient Chinese architecture from different perspectives. However, some of these opinions inevitably involve subjective speculation. The formation of structural forms is a result of a long history of development and evolution, and it can not simply be attributed to a single direct cause. Currently, there is no conclusive evidence to support any particular theory. From the structural practices of ancient architecture, we can infer that curved lines and other structural features were deliberately implemented under strict technical specifications. They are not accidental or the result of technical deficiencies. Based on the evidence from Han Dynasty remains, such as monumental arches, pottery houses, stone shrines, and Han portraits, the roofs of Han Dynasty buildings were mostly straight. Some scholars have inferred that curved roofs may have appeared after the Six Dynasties period. Descriptions of houses in Han Dynasty literature also express the "reverse vault" of curved roofs, and there are also examples of curved roofs in physical artifacts. It is possible that the predominantly straight roofs of Han Dynasty stone vaults were influenced by production techniques and material limitations, or they may be a simplified representation.

In fact, in order to delve deeper into the underlying reasons, it is necessary to analyze each specific feature in detail. The formation of the structural forms of roofs in ancient Chinese architecture is undoubtedly the result of multiple factors and a combination of various conditions.

Overhanging eaves. In ancient architecture, the walls were often made of rammed earth or wooden partitions. In order to prevent erosion from rain and snow, the eaves of the houses were extended to block wind and rain. This became a characteristic feature of ancient architectural roofs. There are two methods

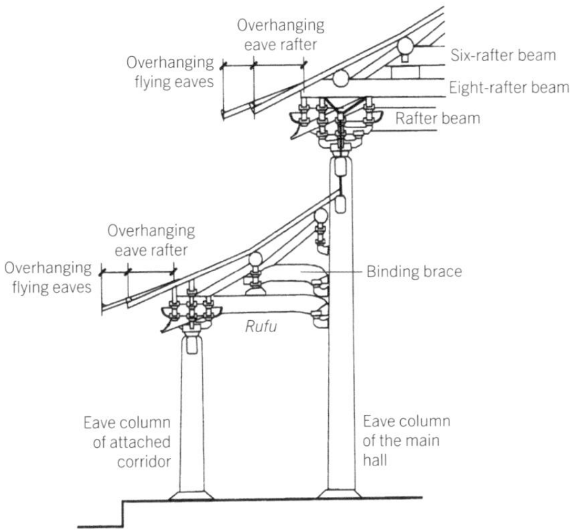

Overhanging eaves in *Yingzao Fashi*

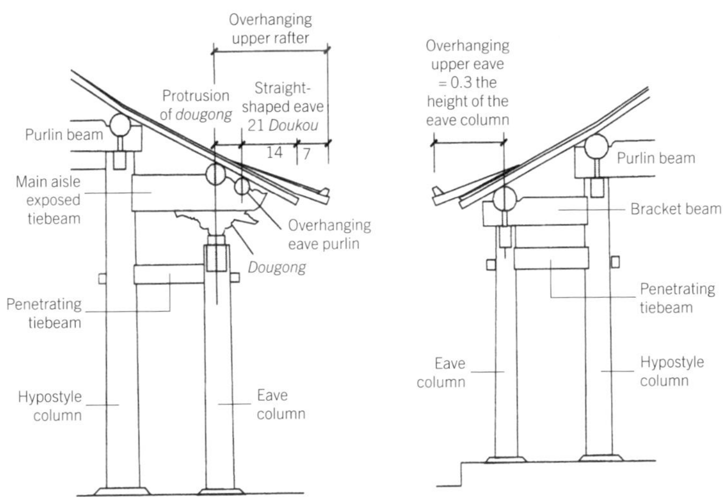

(a) Overhanging upper eave with *dougong*

(b) Overhanging upper eave without *dougong*

Overhanging upper eaves in *Gongcheng Zuofa Zeli*

Methods of overhanging eaves in the Song and Qing dynasties.

of overhanging eaves: with or without *dougong*. Although the proportions may vary, the depth of the overhanging is closely related to the heights of the eave columns. Except for small-scale buildings, it is common to add flying brackets (*feizi*, or flying rafters) on the eave purlins. The upwardly curved flying eaves, with their rear parts cut into sloping surfaces and placed on the eave purlins, not only serve as a buffer for rain and snow but also allow more light into the interior. This design also enhances the upward curve of the roof, creating a sense of lightness in its appearance. In many regions today, it is still a common practice to add diagonal braces for rain protection on roofs without overhanging eaves. It is unclear whether the practice of adding diagonal braces to the front eaves was influenced by the overhanging eaves or vice versa. However, it is more logical to understand that the initial purpose was to protect against rain and snow, with separate rain protection structures added on the front eaves, which gradually evolved to become integrated with the roof. The rain canopy not only protects against rain and snow, but also allows light to enter the interior. The upward inclination of the rain canopy is more conducive to balancing both aspects. The diagonal beams merge with the roof structure and gradually evolve into *dougong*. In this way, the upward curve of the roof with the flying eaves becomes more understandable.

Roof curves. The roof curves in multiple directions, including the eave curve formed by the upward bend at the ends of the eave lines, the ridge curve formed by the upward bend at the ends of the roof ridge, and the roof curve formed by the roof truss.

The formation of the eave curve is structurally due to the rise of the eave columns. Additional components such as corner beams are used to make the corners of the roof higher, and *shengtoumu* (upward-tilting wood) may also be used

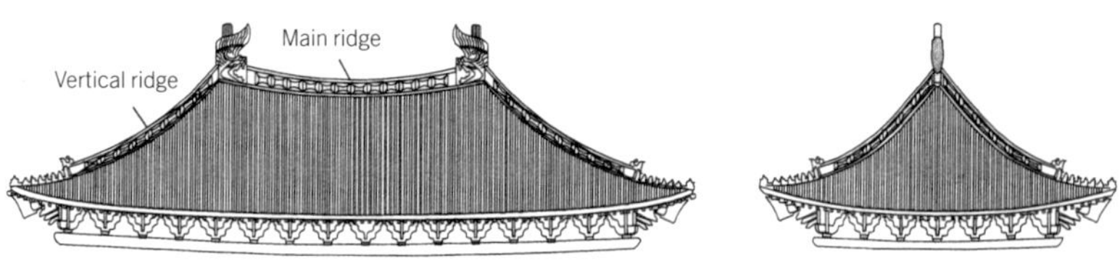

Roof curves.

under the eave purlins. It is difficult to determine whether eaves were straight or curved prior to the Tang Dynasty due to the lack of surviving wooden structures. Murals, stone architecture, and *mingqi* (funerary objects) mostly show buildings with straight eaves, but there are also some architectural depictions with curved eaves. Wooden architecture in the Tang Dynasty had both straight and curved eaves. However, our understanding of eaves in Tang and earlier architecture has come mainly from materials such as the Dunhuang murals, which mostly depict buildings with straight eaves. This information has formed a fixed perception of eaves in Tang and earlier architecture. In addition, the roof is the most vulnerable part of a building and undergoes many repairs. Because of the fixed perception of straight eaves in earlier dynasties, there is no definitive answer as to whether the eaves were originally straight or curved. The practice of restoring the eaves to a straight shape is based on the principle of "repairing as it was,"[9] but it is difficult to determine with certainty.

The curved ridge line was already visible in the Han Dynasty. Architectural depictions in Han Dynasty stone buildings and *mingqi* (funerary objects) show that the upper part of the roof structure was straight, but the ridge line was already rising. Surviving examples of wooden architecture from the Tang, Song, and Yuan dynasties show the use of *shengtoumu* at both ends of the ridge beam, resulting in a noticeable upward curve of the ridge. In the Ming and Qing Dynasty architecture, the ridge line became straight, but in some southern regions, curved ridge lines can still be seen in the Ming and Qing Dynasty buildings.

The roof surface curve can have both longitudinal and transverse curves. From a structural perspective, the transverse curve is formed by the rise of the eave columns and the formation of wing corners. In the longitudinal direction, it is mainly formed by the *juzhe* (rise and fall) (also known as *jujia* in the Qing Dynasty) and *feichuan* (flying purlin). *Juzhe* refers to adjusting the height of the beam's tenon to create a slight concave curve in the entire roof surface. The *juzhe* in the Song Dynasty had a steep slope at the curve of the roof surface, a large overhang, a small slope near the eaves, and a light and elegant appearance, which differed from the Qing Dynasty's *juzhe* with a steeper and gentler curve, a shallow overhang, and a dignified appearance. There were also differences in specific practices. According to the *Yingzao Fashi*, the method of *juzhe* of the roof is described as follows: "The method of folding the roof is to raise it by one unit for every *chi*, and each truss is reduced by half. For example, if the roof is raised two *zhang*, start at the back of the ridge beam and go down to the back of the rafter end at the eaves. The first fold is two *chi*, and starting from the back of the first fold, go down to the back of the rafter end at the eave. The second fold is one *chi*. If there are more folds, continue in the same manner, decreasing by half at each fold, down to the back of the rafter end at the eaves." This method involves determining the height of the ridge as 1/3 of the distance between the front and rear eave purlins. A diagonal line is drawn from the top of the ridge beam to the upper part of the eave purlin. From the top, the first horizontal beam is placed at 1/10 of the ridge height below the line, and then a diagonal line is drawn from the top of the first horizontal beam to the upper part of the eave purlin. Another horizontal beam is placed at 1/20 of the ridge height below the line (each beam is reduced by half), and so on. The *juzhe* in the Song Dynasty decreased from top to bottom, with the decrease in height determined by the distance between different diagonal lines and the upper part of the beam, similar to a graphical method.

In contrast, the Qing-style *jujia* method is based on numerical calculations and involves a gradual increase of the height from bottom to top. It is based on the horizontal distance between the two purlins, which is the basis for the *bujia* (step frame). The height of the *bujia* is determined by the size of the building and the number of purlins. The ridge beam is set at nine *bujia,* which is 9/10 of the total length of the step frame. According to the *Gongcheng Zuofa Zeli*, it

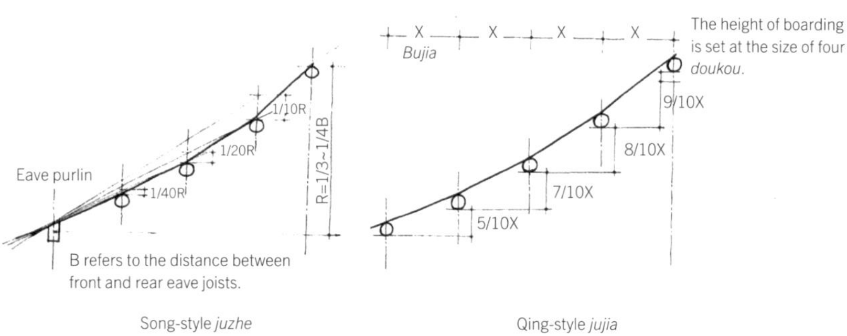

Song-style *juzhe* and Qing-style *jujia*.

states, "For example, if the eave section has five *jujia* (meaning if the horizontal length of the step frame is one *chi*, then the height is five *cun*, and so on), the flying eave has three to five *jujia*, and the ridge of a five-purlin roof has seven *jujia*. For a seven-purlin roof, the ridge has nine *jujia*. For a nine-purlin roof with a lower gold section, it has six and a half *jujia*; the upper gold section has seven and a half *jujia*, and the ridge has nine *jujia*. For an eleven-purlin roof with a lower gold section, it has six *jujia*; the middle gold section has six and a half *jujia*, the upper gold section has seven and a half *jujia*, and the ridge has nine *jujia*. Alternatively, it can be determined based on the specific shape."

Wing corners. It is uncertain whether the wing corners are formed first, leading to the curved eaves and the rise of the eave columns, or whether the curved eaves and the rise of the eave columns come first, resulting in the wing corners. However, regardless of the sequence, the wing corners, curved eaves, and the rise of the eave columns are interrelated. At the intersection of the roof eaves, the corner beams (known as the main corner beam and subsidiary corner beam), which are inclined at a 45-degree angle, are larger and higher than the rafters. To make the rafters and corner beams level, the rafters near the corner beams gradually increase their slope, causing the rafter ends to gradually tilt upward until they are parallel to the angled beams. The upward tilt of the wing corners creates a graceful form, which is a distinctive feature of ancient Chinese architecture and has been depicted in numerous literary works. *The Book of Songs* describes it as follows: "Like the wings of a bird, like the thorns of an arrow, like the skin of a bird, like the soaring flight of a bird, the noble person ascends." The wing corners of houses in southern China are generally more uplifted than those in northern China, with the south being delicate and the north being majestic, which is similar to the character and temperament of people in the north and south. The uplifted wing corners of houses in southern China can be achieved by two methods. The first method is called *shuiqiang faqiang*, in which the wooden frame of the corner itself does not lift. Instead, the ridge beam is used to create the lift, resulting in a curved, upward shape resembling a rolled leaf. The second method is called *nenqiang faqiang*, which is similar to the raised wing corners of northern China, but with slight differences in construction. In this method, the subsidiary corner beams are not attached to the upper part of the main corner beam, but rather at the end of the main corner beam.

Wing corners.

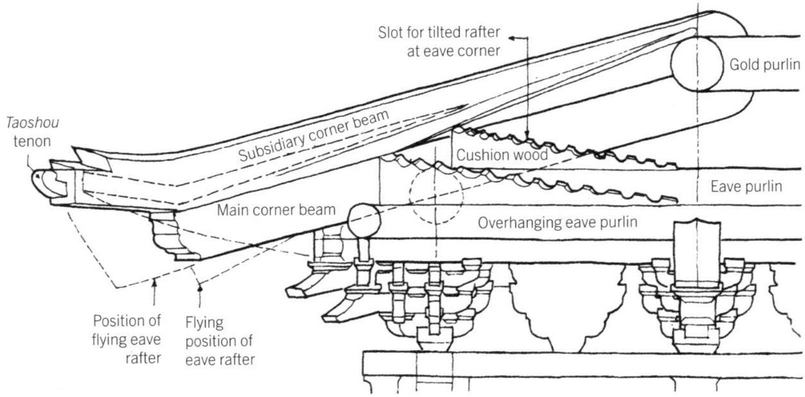

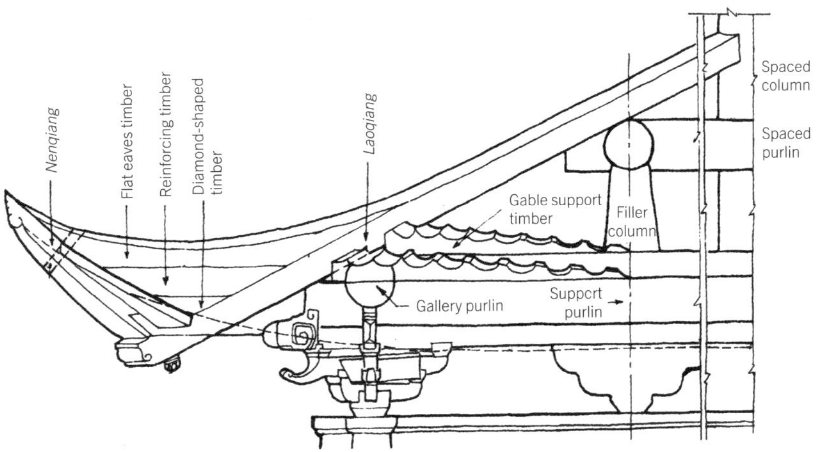

TOP Qing-style wing corner.

BOTTOM The *nenqiang faqiang* of the wing corner in the Jiangnan region.

Ridge

In ancient Chinese architecture, the roofs are sloped, and the intersections and joints of the roof are prone to water leakage. This applies to the intersections of roof slopes, as well as the intersections between the roof and the gable walls. Structural treatments are applied in these areas to prevent water leakage while also considering the aesthetic aspect. This gives rise to the concept of the ridge in Chinese architecture. The ridge can be classified into different types according to its location and function, such as the main ridge, vertical ridge, diagonal ridge, surrounding ridge, and horizontal ridge. These ridges not only serve as waterproofing elements but also become important decorative features in ancient Chinese architecture, showing the unique charm of traditional buildings.

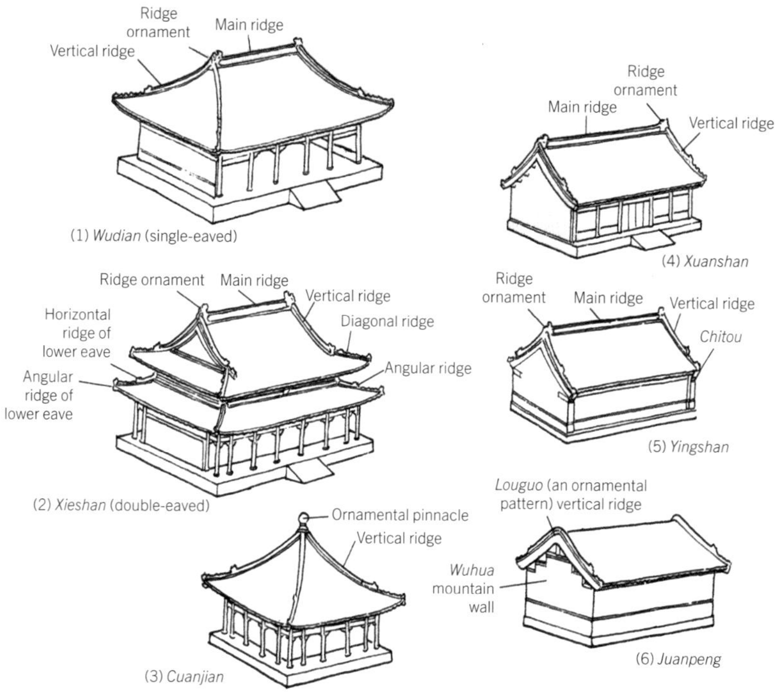

Ridges in ancient Chinese architecture.

Decorations of the main ridge.

Main ridge. The main ridge consists of a long ridge body and components at both ends. It is located at the top intersection of the roof and serves as the boundary of the highest point on the front facade of the building. For *wudian* roof, the main ridge is typically in a pointed style, while for *yingshan*, *xuanshan*, and *xieshan* roofs, there are two types of main ridges: pointed style and curved canopy style. The materials used for the ridge body in ancient times were often roof tiles, which were sometimes coated with mortar or crafted into hollow patterns to facilitate ventilation. Later, specialized kiln-fired ridge components were gradually introduced. The central upper part of the main ridge is often decorated with images of mythical creatures such as the vermilion bird or the black turtle. In temples and Taoist temples, there may be precious ornaments or fireballs, with two flames on either side, accompanied by carved dragons or animal faces, forming a double dragon and pearl motif. The main ridge may also be adorned with decorative patterns on the sides.

LEFT Main ridges and *chiwen*.

RIGHT Decorations of the main ridges.

Chiwen.

According to the *Yingzao Fashi* of the Song Dynasty, the decoration at the ends of the main ridge is typically done with *chiwen*. The *Tan Bin Lu** records, "In the East Sea, there is a fish-like creature called 'qiu,' with a tail resembling a *chi*. When it flaps its tail, rain falls. Therefore, it is placed on the roof as a symbol of auspiciousness for rain prevention and fire extinguishing." The *chiwen* came from *chiwei* and can be traced back to architectural ornaments of the Han Dynasty. Clear depictions of *chiwei* can be found in the murals of Jiuyuanguan during the Eastern Wei and Northern Qi periods. The earliest intact artifact resembling a *chiwei* was unearthed at the Yecheng site in Hebei Province. *Chiwen* images can also be found in stone caves and tombs from the Sui and Tang dynasties. In the middle to late Tang and Liao dynasties, the lower part of the *chiwei* began to feature an open-mouthed animal head, which gradually evolved into a fishtail. During the Song Dynasty, there were several forms of *chiwei*, including *chi* tails, dragon tails, and animal heads. In the Yuan Dynasty, *chiwei* gradually curved outward, and some had already transformed into *chiwen*. In the Ming and Qing dynasties, the *chiwen*'s tail completely curved outward, with the end transitioning from a forked shape to a curved shape. The body of the creature was intricately carved, often accompanied by small dragons, and had a square shape. There was a sword handle on its back, and its name changed to *shouwen* or *dawen*.[10]

* A books records anecdotes of the Tang Dynasty.

Chiwen.

The use of *chiwen* is generally reserved for higher-level buildings, while lower-level buildings may use *wangshou* or no decorative elements at all. The presence or absence of chiwen is also a distinguishing factor between palace buildings and non-palace buildings. In the Jiangnan region during the Ming and Qing dynasties, *puwa zhuji* (tile ridge construction) was commonly used. Large buildings used glazed tiles, while small buildings used clay tiles. In some folk buildings in the Jiangnan region, only the main ridge was installed without using vertical ridges. The main ridge can take various forms, including *youji* (wandering ridge), *ganzheji* (sugarcane ridge), *wentouji* (patterned ridge), *bujiji* (chicken feeding ridge), *cimaoji* (female hair ridge), *bulongji* (feeding dragon ridge), *pitiaoji* (leather strip ridge), and *biandanji* (bamboo pole ridge). The chicken and dragon heads in the *bujiji* and *bulongji* face outward, and the upward-curving tails are painted and made of bent iron plates.

Vertical ridge. The vertical ridge is located at the eaves of the side walls (or at the intersection of the roof in the case of *wudian* and *cuanjian* roofs) and is perpendicular to the main ridge. It is used in various types of roofs, including *yingshan*, *xuanshan*, *xieshan*, *wuxian*, and *cuanjian* roofs. The vertical ridge is divided into two main sections: the front section, known as *chuishouqian*, and the rear section, known as *chuishouhou*. The rear section of the vertical ridge is placed on the framework structure, while the front section is placed on the subsidiary corner beam of the wing corner. To prevent the sub-rafter from being exposed to rainwater erosion, a decorative animal figure, known as *taoshou*, is often used on the sub-rafter. According to the *Yingzao Fashi* of the Song Dynasty, the

Chiwen.

vertical ridge of official-style buildings is adorned with decorative animal figures, known as *chuishou*. At the end of the ridge, a figure called *pingjia* is used (in Buddhism, it refers to a beautiful bird, often depicted as a bird-bodied celestial being, later associated with heavenly deities and warriors; in the Qing Dynasty, it was depicted as an immortal). The *pingjia* is followed by 2 to 9 squatting animals (they were depicted as walking animals in the Qing Dynasty). For non-palace buildings without eaves at both ends, only one *pingjia* or squatting animal is used. In the Qing Dynasty, the walking animals were usually depicted as nine, but in the Hall of Supreme Harmony in the Beijing Forbidden City, there are ten. After the figure riding a phoenix, the sequence is: walking dragon, flying phoenix, walking lion, celestial horse, seahorse, *suanni* (scaled lion), *yayu* (playful fish), *xiezhi* (a mythical creature), *douniu* (fighting bull), and *hangshi* (a mythical creature). The number of beasts used depends on the length of the roof slope and the height of the eaves. Generally, one walking animal is used for every two feet of column height, and odd numbers are preferred.

In addition to the main ridge and vertical ridge, there are other types of ridges used in traditional architecture. The diagonal ridge is found in *xieshan* roofs, where the ridge extends diagonally from the corners of the roof at a 45-degree angle. It is also used in double-eave roofs where the ridge extends from the eaves at a 45-degree angle. This is an important distinction from the vertical ridge of palace roofs. The diagonal ridge is also divided into front and back sections, similar to the vertical ridge. The use of diagonal ridges can be traced back to the Han Dynasty. The horizontal ridge is found in the decorative flower-like elements on the hipped roof, connecting to the wind boards. It runs parallel to the slope of the roof. In the Song Dynasty, it was referred to as *quji*. These additional ridges add decorative elements and architectural details to the roof, enhancing its aesthetic appeal. The *paishan* ridge is a collective term for the vertical ridges used in small buildings with *xieshan*, *yingshan*, and *xuanshan* roofs. *Paishan* literally means to follow the shape of a mountain, indicating that the ridge hangs down along the peak of the roof, as opposed to the diagonal direction of the vertical ridge in palace roofs. The decorative ends of the *paishan* ridge are called *paishan goudi*, which refers to the decorative hooks and water drips. In addition, due to variations in architectural styles, time periods, and regional differences, ancient buildings have different types of ridges and more complex names, depending on the complexity of the roof structure. However, the main types of ridges include the main ridge, vertical ridge, diagonal ridge, and horizontal ridge.

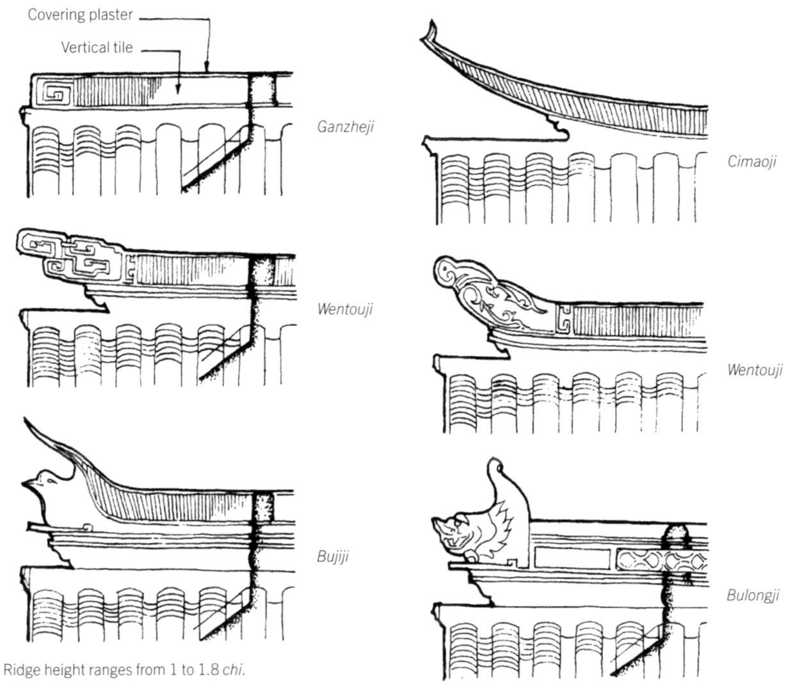

Suzhou-style main ridges.

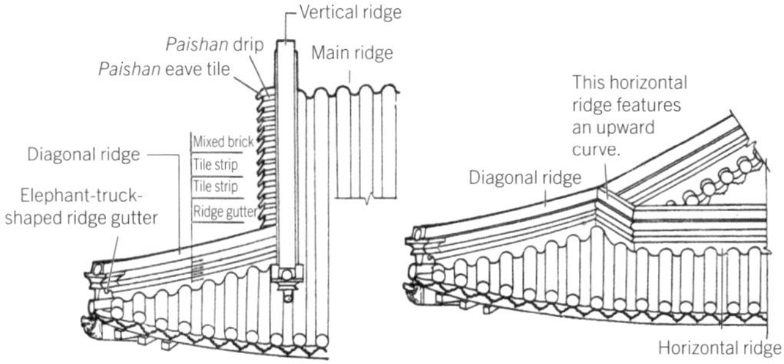

Vertical and diagonal ridges of small *xieshan* roofs

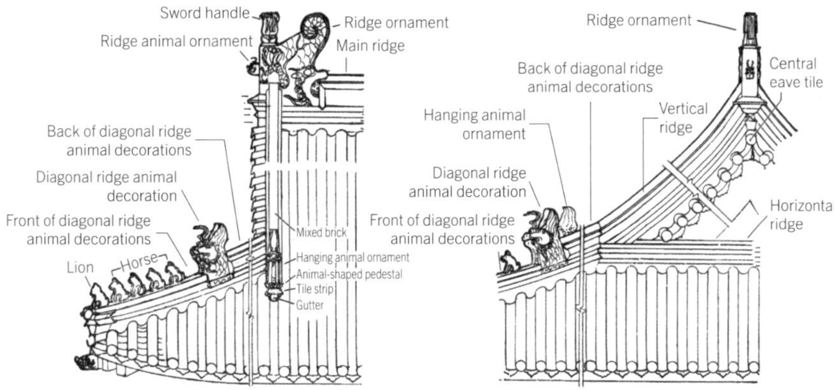

Vertical ridges, diagonal ridges, hanging animal ornaments, and diagonal ridge animal decorations of large *xieshan* roofs

Vertical ridges and hanging animal ornaments of *xieshan* roofs.

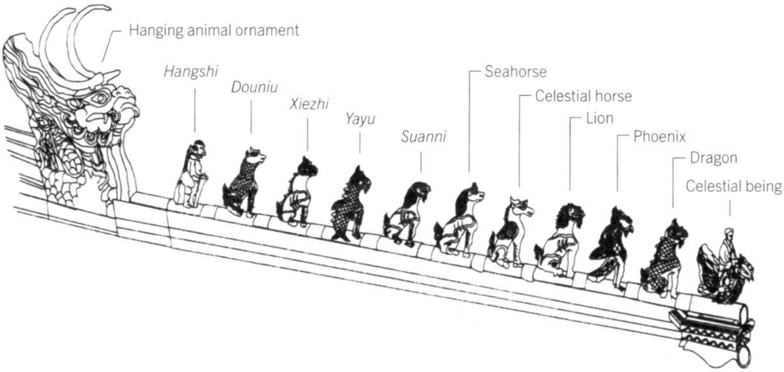

TOP Celestial being and walking animals in the Qing Dynasty.

BOTTOM *Pinjia*.

TOP LEFT Hanging animal ornaments and *taoshou*.

BOTTOM LEFT & TOP RIGHT *Pinjia* and *taoshou*.

Hanging animal ornaments.

Vertical ridge
Diagonal ridge
Horizontal ridge

Hanging animal ornaments.

NOTES

1. Liu Zhiping, *Chinese Architectural Types and Structures* (Beijing: China Architecture & Building Press, 1987), 69.
2. Pan Guxi, *The History of Chinese Architecture* (Beijing: China Architecture & Building Press, 2015), 284.
3. Ibid., 287.
4. Star tiles are tiles with nail fixing devices. They are used at appropriate intervals on the roof ridge to secure the entire roof ridge.
5. Liang Congjie, *Lin Huiyin's Collected Works of Architecture* (Beijing: Baihua Literature and Art Publishing House, 1999), 8.
6. Le Jiazao, *The History of Chinese Architecture* (Beijing: China Ancient Books Publishing House, 2016), 17.
7. Liu Zhiping, *Chinese Architectural Types and Structures* (Beijing: China Architecture & Building Press, 1987), 18.
8. Joseph Needham, "Civil Engineering and Nautical Technology," in *Science and Civilisation in China*, vol. 4 (Beijing: Science Press, 2008), 96.
9. The main hall of Nanchan Temple in Wutai Mountain, Shanxi Province, was built in the third year of Tang Jianzhong's reign (AD 782). The roof had been repaired during the Qing Dynasty. During the renovation in the 1970s, the roof was reconstructed to restore its historical appearance, following the architectural forms depicted in the mural of Dunhuang Grottoes, with the corners of the eaves kept as straight as possible.
10. Pan Guxi, *The History of Chinese Architecture* (Beijing: China Architecture & Building Press, 2015), 288.

CHAPTER 7

JOINERY WORK RENOVATION

Ancient Chinese timber-frame architecture, after the completion of the enclosing structure of the foundation, beams, walls, and roof, requires further refinement in the form of decoration, also known as joinery work (*xiaomuzuo,* so named because they are small woodwork compared to columns and beams, which are considered large woodwork, i.e., *damuzuo* [carpentry work]) in the Song Dynasty.

Decoration can be divided into exterior decoration with railings, doors, windows, etc., and interior decoration with ceilings, caisson ceilings, and partitions.

Gates and Doors

Doors are functional passages used in houses, city walls, intersections, streets, courtyards, and other entrances and exits, serving as junctions between adjacent spaces. According to the *Explanation of Terms* written by Liu Xi during the Eastern Han Dynasty, "门 (mén), means to touch, serving as the tactile curtain for protection; 户 (hù), means to guard, ensuring careful protection and closure." In the *Jade Chapter* by Gu Yewang of Southern Dynasties, a lexicon book compiled by Gu Yewang during the Southern Dynasties, it is stated: "In the main hall, it is called 户 (hù), in the defined area, it is called 门 (mén)." In ancient Chinese architecture, the term "门" (mén) carries two distinct meanings: one as the opening and closing passage or gate connecting adjacent spaces, such as the doors of houses, courtyards, city gates, etc.; the other referring to the door itself, specifically the door leaf, as mentioned by Gu Yewang, known as "户" (hù). The *hu* and *men* in ancient Chinese architecture refer to doors and windows. In *Huainanzi · Shuolin Xun*, it is noted, "The illumination from ten 牖 (yǒu) cannot be compared to the brightness of a single 户 (hù)," meaning that the radiance brought forth by the opening of ten windows cannot be compared to that produced by the opening of a single door. In early times, the term "户牖" (hù yǒu) referred to both doors and windows.

Doors hold an important place in ancient Chinese architecture. As mentioned in the previous chapter, the front facade of ancient Chinese buildings often incorporates partition screens or a combination of partition screens and window sills to form a protective enclosure for the front eaves. This kind of

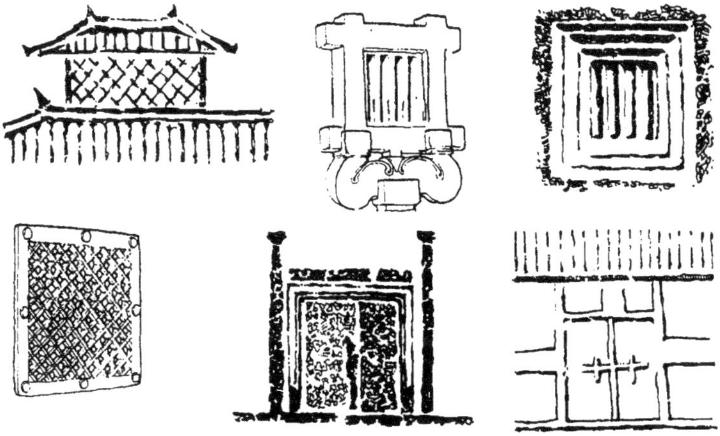

Doors and windows in the Han Dynasty.

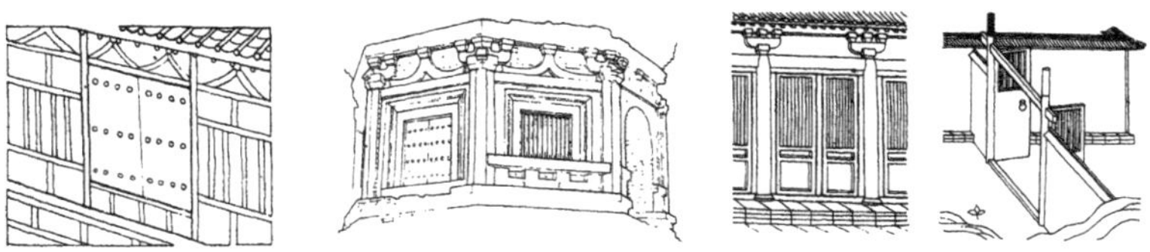

Doors and windows in the Southern and Northern and Tang dynasties.

Doors and windows in the Song, the Liao, and the Jin dynasties.

enclosure not only considers the ventilation and lighting functions of the house, but also fully reflects the visual requirements of the transition between the interior and exterior of Chinese architectural courtyards. Partition doors have become the inner walls of the courtyards, which is also one of the characteristics of ancient Chinese architecture.

Doors have an important ceremonial function in ancient Chinese architecture, and the main gate of a courtyard is a symbol of social status. Doors serve as significant symbols of identity, wealth, and social status, with terms like "朱门" (zhū mén: red gate, symbolizing a wealthy and influential family), "豪门" (háo mén: noble family or wealthy family), "侯门" (hóu mén: literally "Marquis gate," referring to a noble family), "寒门" (hán mén: literally "cold gate," referring to a humble family or poor family), and "柴门" (chái mén: literally "thatched door," symbolizing a humble or rustic family) denoting various levels of status and wealth. A door alone can distinguish between rulers and subjects, the distinguished and the common, delineating distinctions of privilege and rank. In *The Rites of Zhou*, the regulations for the gates of the imperial palace are outlined as follows: "The king has five gates, namely the Outer Gate (*gaomen*), the Pheasant Gate (*zhimen*), the Treasury Gate (*kumen*), the Responding Gate (*yingmen*), and the Road Gate (*lumen*)." This means that the imperial palace can have five layers of gates, while the vassal lords are only allowed to construct three layers of gates, namely the Treasury Gate, the Pheasant Gate, and the Road Gate. This system, known as the "Five Gates Protocol," has been followed by successive emperors for thousands of years. In the Ming and Qing dynasties, the Forbidden City, the imperial palace complex, had five layers of gates from the Great Gate of the Imperial City to the front of the Hall of Supreme Harmony: Great Qing Gate (*gaomen)*, Tian'anmen (*zhimen*), Duan Gate (*kumen*), Meridian Gate (*yingmen*), and Gate of Supreme Harmony (*lumen*). In the Han Dynasty, the neighborhood system (*lifang* system) regulated the classification of residences. According to the *Memorial of the King of Wei*, "If the entrance of a house faces the main road and does not go through the inner lanes, it is called *di*. Feudal lords with fiefs of less than ten thousand households are not allowed to be called *di*. Those who live within the inner lanes, regardless of their status, are not called *di*." The residences of feudal lords, high-ranking officials, and those with fiefs of ten thousand households or more are called *di*. They can have their entrances directly facing the main road and do not need to go through the inner lanes. On the other hand, the residences of those with fiefs of less than ten thousand households are called *she*, and they must go through the inner lanes. The Tang Dynasty also had regulations on the neighborhood residence system. According to the record in Volume 86 of *Tang Huiyao*, in July of the fifth year of the Taihe era, an official report stated, "Those below the rank of third class, as well as those in the neighborhood who do not possess the three qualifications, should not arbitrarily open doors to the street at their convenience without any restrictions. Due to long-standing practices, it has become difficult to enforce regulations. Sometimes, doors are opened before the drum is beaten or even late at night without any sound. This makes it difficult for the street inspectors to cover the area, and it also facilitates the escape of criminals… If they do not have the three qualifications, please restrict them from opening doors in the neighborhood and seal off all street entrances." This means that high-ranking officials below the rank of third class and those without special qualifications were not allowed to open doors facing the street on the neighborhood walls. In Bai Juyi's poem "Shang Zhai," he asks, "Whose home has a grand mansion with a vermilion gate by the main street?" And he answers himself, "The master sits inside; after ten years, he becomes a high-ranking official." The vermilion gate of the grand mansion faces the bustling main street, which represents the privilege of the gate.

The Ming and Qing dynasties had even stricter and more detailed regulations on door decorations, including paint, nails, and rings. The book *Ming Shi* (*History of the Ming Dynasty*) records similar regulations, "The main gate of the princess's residence has five rooms and seven rafters, with green oil, copper rings, stone foundation, and engraved bricks with exquisite patterns. The gates of dukes' residences have three rooms and five rafters, with gold paint and tin rings with animal faces. Officials of the first and second ranks have three rooms and five rafters, with green oil and tin rings with animal faces. Third to fifth rank officials have three rooms and three rafters with black oil and tin rings. Sixth to ninth rank officials have one room and three rafters, with black doors and iron rings." In the Qing Dynasty, there were detailed regulations regarding the use of door bolts and decorative cylinders. According to the *Daqing Huidian* (*Collected Statutes of the Great Qing*), "The doors of palace buildings are built on elevated foundations, covered with yellow glazed tiles, and adorned with golden door studs." "At the four gates of altars, temples, and circular mounds, both inside and outside the walls, there are vermilion doors with golden door studs, arranged in a grid of nine by nine." The number of doorknobs used in the residences of princes, dukes, and marquises was specifically defined, "The residence of a prince has a main gate with five rooms, with nine vertical door studs and seven horizontal door studs." "The residence of a crown prince, with the same number of rooms, has door studs reduced by seven divided by two (subtracting two-sevenths)." "The residences of dukes, *beile* (prince of the second rank), *beizi* (prince of the third rank), *zhenguo gong* (duke who stabilizes the nation), and *fuguo gong* (duke who assists the nation) are the same as those of princes." "The door studs in the residences of commoners are all seven in both vertical and horizontal directions, gradually decreasing to five by five, all made of iron." The gates of the imperial palace are embellished with eighty-one door studs, those of the residences of princes with sixty-three, and those of the residences of dukes and marquises with forty-nine, and those of officials with twenty-five door studs. The houses of common people have no nails, so the common people are called *baidinger* (people without nails). In the Tang and Song dynasties, the doors of ordinary people were not painted and were left in their original wood color, and they were called *baibanfei* (white-board doors). The Tang Dynasty poet Wang Wei wrote in his poem "Farmer's House": "Birds nestle eggs by the well covered in green moss, while roosters crow in front of the *baibanfei*." The poet Dai Fugu of the Southern Song Dynasty wrote in his poem "Night Stay at a Farmer's House": "Knocking on the unpainted door of a farmer's house at night." In the novel *Jin Ping Mei* (*The Plum in the Golden Vase*), it is mentioned in the 72nd chapter, "Lady Li dreams of the house," and when Ximen Qing passed by Zao Fu Alley, he saw "a pair of unpainted doors." To lock the central threshold and the connected lintel, wealthy households use four, smaller households use three, while the gates of the royal family and palaces may use up to twelve door cylinders. It can be seen that all components related to doors are expressions of ceremonial hierarchy.

Doors in ancient Chinese architecture also have important *feng shui* functions. Doors are considered the *qi* (energy) mouth of a house in *feng shui*, and there are many taboos in folk customs regarding their placement and design. Generally, the main entrance of a house is located in the southeast corner of the yard, and the main house facing north and south has three auspicious positions in *feng shui*: *li* (south), *xun* (southeast), and *zhen* (east). Among them, the southeast direction is known as the *Qinglong* Gate and is considered as the best direction. It is also common to have larger courtyards and multiple courtyards, with the rooms located at the back. The auspicious orientation is in the *li* position (south), and the main gate should not face the street or sharp corners to avoid *sha qi* (negative energy). In addition, there should not be a pond or large trees planted in front of the main gate, as these are *feng shui* taboos that have some connection to environmental science.

Doors and gates can be divided into different types according to their location and purpose, such as room doors, house gates, courtyard gates, garden gates, lane gates, city gates, temple gates, etc. During the Ming and Qing dynasties, the courtyard houses in the northern region had various types of entrance gates, such as house-style gates, wide and bright gates, golden pillar gates, *manzi* gates, *ruyi* gates, and hanging flower gates. Based on the design of the gates themselves, there were panel gates, partition gates, solid threshold gates, checkerboard gates, and more. In the Song Dynasty, there were also plate gates, soft gates with grids, and *wutou* gates.

***Heng* gate and *wutou* gate.** *Heng* gate was the early form of courtyard gates. *The Book of Songs · Chen Feng* describes, "Below the *heng* gate, one can rest at ease." The term *heng* here refers to the horizontal structure above the pillar heads on both sides of the entrance, where one or two long horizontal beams are placed. This is the simplest and most primitive form of a gate. This type of gate appears frequently in Chinese paintings throughout history, and similar gates can still be found in some remote rural areas. *Heng* gates later evolved into *wutou* gates, which were used in the residences of the noble class during the Sui and Tang dynasties. In the *Tang Liudian* (*Six Codes of Tang Dynasty*), it is recorded, "For officials of the sixth class or higher, the *wutou* gate is still commonly used." In the murals of the early Tang Dynasty, one can see the shape

Wutou gate in the *Yingzao Fashi*.

of the *wutou* gate, such as the murals of Cave 98 in the Dunhuang Grottoes. The *wutou* gate is constructed with two round columns supporting a horizontal beam, with the ends of the beam uplifted. The tops of the columns are covered with black ceramic column sleeves. According to the *Cefu Yuangui*,* it states, "Tiles are placed on the ends of the pillars and dyed with black ink, which is called *wutou* (black-headed) dye." This is also the origin of the name *wutou* gate. The upper part of the *wutou* gate is fitted with straight latticed windows. The pillar on the left side of the gate is called *fa*, symbolizing accomplishments, while the one on the right side is called *yue*, representing long-standing experience and generations of officials holding high positions. If the gate is wide, two additional pillars can be added, as recorded in the *Yingzao Fashi*. *Wutou* gate later evolved into the lintel-style wooden or stone gate located along the central axis of Confucian temples.

***Ban* gate (panel gate).** Panel gate (*ban* gate) (also known as the Song-style *ban* gate) has been used since ancient times, appearing on bronze artifacts from the Zhou Dynasty and stone portraits from the Han Dynasty. Existing examples include the gate of Foguang Temple in Wutai Mountain. *Ban* gate is widely used as the main entrance for city gates, palaces, government offices, temples, and residences. *Ban* gate can be further divided into the checkerboard and mirror panel. Checkerboards are made with large edges and lintels, and then door panels (elbow panels) are installed between the lintels using several transverse belts, resembling a checkerboard. Some gates are tightly connected with iron nails, and the heads of the nails are decorated with auspicious patterns. Gates without any decorative lines, but with a smooth surface like a mirror, are called mirror panel gates. Furthermore, there exists another variant of the door leaf, crafted from thick wooden panels and assembled using door hinges as connecting elements. One end of the door hinge is inserted into the wider edge of the attached door pivot, without using lintels above and below, called *shita* gate (solid threshold gate). The *Yingzao Fashi* also records a type of soft gate, which evolved from the panel gate but is lighter in form, enclosing a wooden frame adorned with inset thin panels. The Song Dynasty produced variations like the tooth-guard-seam soft gate and the combined-panel soft gate. The design of the soft gate is

---

* The largest encyclopedia compiled during the Song Dynasty (AD 960–1279).

characterized by its flexibility, allowing for a variety of forms. The concept of partition gates, in particular, represents an evolution that builds based on the foundation of soft gates.

Ban gate of Foguang Temple in Wutai Mountain.

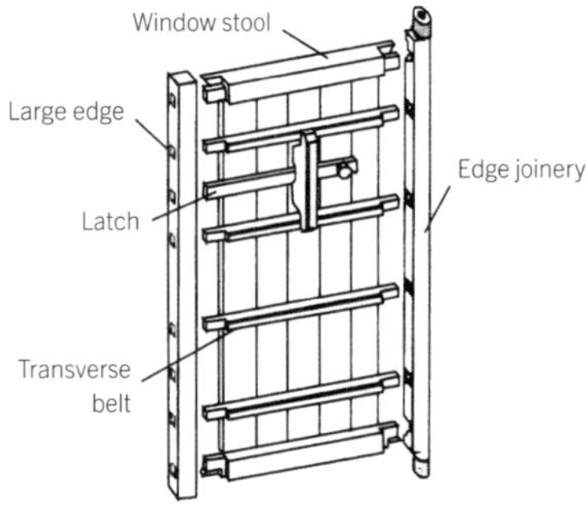

Edge-joined window and door

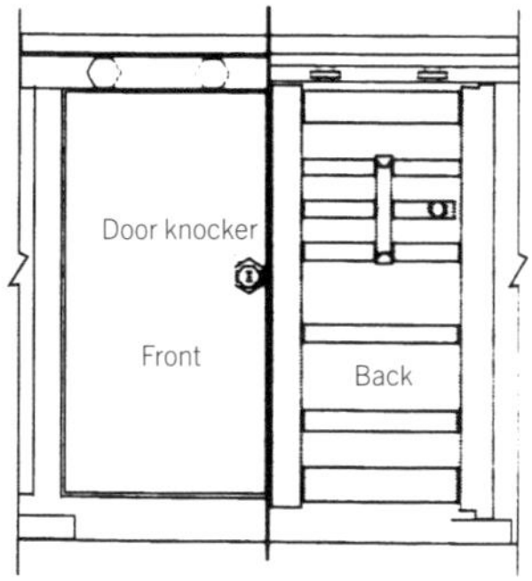

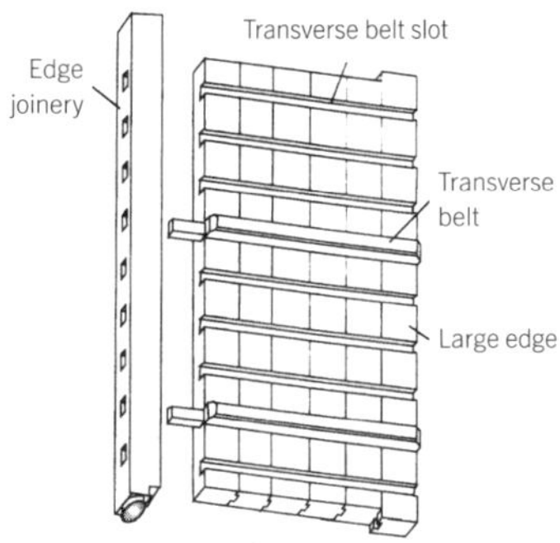

Shita door and window

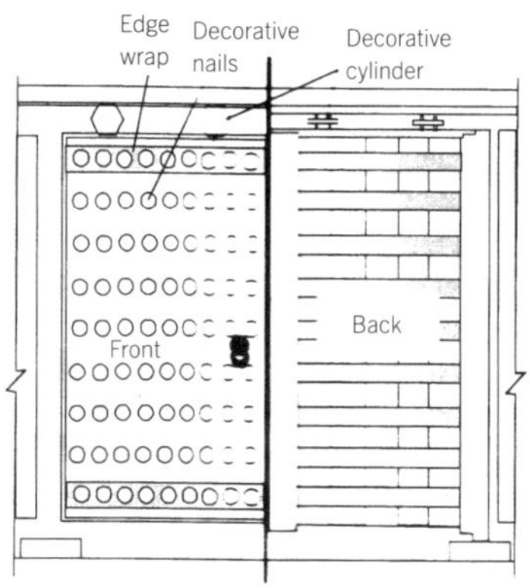

TOP Chessboard-style paneled gate.

BOTTOM Solid threshold gates.

TOP *Shita* gate.

MIDDLE Exterior and interior views of the *shita* gate.

BOTTOM *Shita* gate in residential housing of the Yuan Dynasty.

Partition gate. Partition gates, also known as *geshan* gates in the Song Dynasty, are characterized by their lightweight frame and are commonly used for interior and exterior doors of individual buildings. Partition gates are installed within the threshold frame, with four, six, or eight partitions per door. The width-to-height ratio of each partition is about 1:3 to 1:4. Each partition can be roughly divided into an upper and a lower part. The upper part is called *gexin* (referred to as *xinti* or *huaxin* in the Qing Dynasty), and the lower part includes *yaohuaban* (waist panel) (referred to as *taohuanban* in the Qing Dynasty) and the *zhangshuiban* (water-blocking panel) (referred to as *qunban* in the Qing Dynasty). The height of the upper and lower parts follows the four-six ratio, also known as the "four-six partition." The structure includes *menting* (door frames, called *bianting* or *biankaung* in the Qing Dynasty), upper, middle, and lower window stools (the middle window stool is called *yaohcuan*, "waist beads"), as well as the core, waist panel, and water-blocking panel. In the early days, window stools were used less frequently. In the Tang Dynasty, three window stools appeared. Four window stools were generally used in the Song and Jin dynasties, while five or six were commonly used in the Ming and Qing dynasties. Some corners were reinforced and decorated with copper leaf.

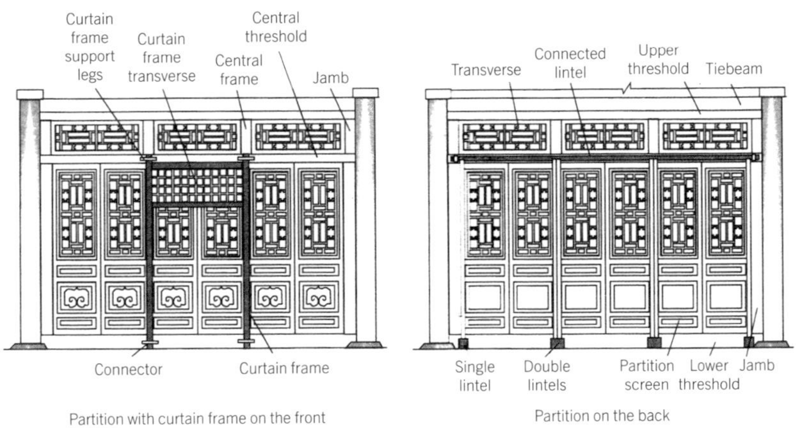

Partition gate.

TOP Partition gate.

MIDDLE Partition gate and sill wall window.

BOTTOM Construction of a single partition panel.

Forms of the partition screens

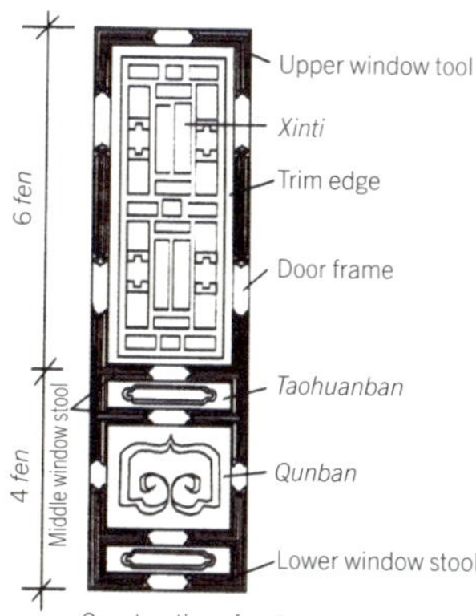

Construction of a single panel

The *gexin*, *yaohuaban*, and *zhangshuiban* of partition gates are the focus of decoration and are often carved. In the Tang Dynasty, *yaohuaban* and *zhangshuiban* were mostly plain, while in the Song Dynasty and later, they were often decorated with carvings of flowers, figures, and other motifs. In the Tang Dynasty, the *gexin* was mostly straight or square, and in the Song Dynasty and later, the patterns became more diverse, becoming even more abundant in the Ming and Qing dynasties.

Official-style buildings in the Song Dynasty often used ball patterns and willow patterns, while the commonly used patterns in the Ming and Qing dynasties include patterns of characters of "万" (*wanziwen*), lanterns, turtle shells (*guibeijin*), interconnected squares (*bubujin*), crutches (*guaizijin*), cracked ice, interconnected circles (*panchangjin*), diamond flowers (*linghuajin*), etc. The more advanced official-style buildings used diamond flower patterns, which can be divided into two types: triple-intersecting hexa-diamond flower patterns and double-intersecting quadra-diamond flower patterns. In the panels of the partition doors, translucent paper, thin gauze, or polished seashells are used to create the "lucky heart" pattern. This design allows light to pass through while blocking wind, dust, and insects, providing privacy and protection.

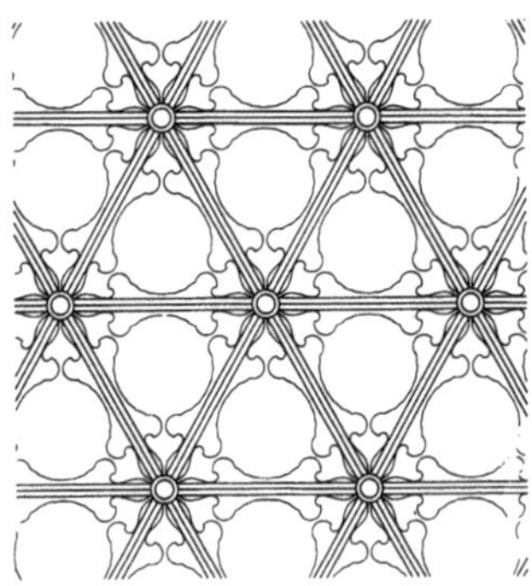

LEFT Commonly used partition gate.

RIGHT Triple-intersecting hexa-diamond flower pattern.

Double-intersecting quadra-diamond flower pattern.

In the later stages of development, partitions had a profound influence on other joinery components. For instance, screens evolved from partition gates. Another type of partition gates that does not reach the floor is called a sill window. Essentially, a sill window is a partition gate with a lower threshold wall without water blocking board. It has a very similar appearance to a regular partition gate.

Windows

Windows in architecture not only provide ventilation, light and view, but also contribute to the overall aesthetic appeal of a building. The size and design of windows naturally receive significant attention in architectural design. Chinese literature is replete with descriptions of windows. "The window frames the snow-capped Western Ridge, while the door anchors the ship sailing thousands of miles in the Eastern Sky." "The sky is tranquil, and the autumn mountains are beautiful. With the window open, the morning light shines through, casting a refreshing green hue." "When will we trim the candle by the western window together and talk about the night rain in *Bashan*." "The intersecting windows form a harmonious scene, like the wings of a phoenix hanging down from the double gates." "The lattice windows are intricately intertwined, and the terrace has three levels of steps." These are just a few examples of the rich descriptions of windows in Chinese literature.

According to *Shuowen*, a lexical book of ancient times, it states, "Inside the house, it is called *cong*, while on the wall, it is called *you*." This translates to: "Inside the house, the window is called the chimney (*cong*), while on the wall, it is called the window (*you*)." As early as primitive societies, windows were used for ventilation, smoke exhaust, and lighting. The upper part of the roof was called *cong* (presumably because in primitive society, the roof and the walls were integrated and later gradually differentiated), while the windows on the walls were called *you*. *You* gradually established itself as the roof and walls differentiated, derived from the transformation of *cong*. In the *Huainanzi · Shuolin Xun*, it is mentioned that "the opening of ten windows is not as bright as the opening of one gate." This implies that the size of the window openings may be relatively small. In fact, even today, some residential buildings in certain regions still use this type of ventilation window on the gable walls. Over time, these windows have evolved into various forms, which differ from the concept of a window that we are familiar with today.

Before the Tang and Song dynasties, windows were mostly fixed and could not be opened. The patterns were characterized by a straight frame resembling an inverted bowl. This type of window is usually fixed and cannot be opened. Straight lattice windows were more common before the Tang and Song dynasties but were gradually replaced by opening windows. In the Ming and Qing dynasties, more complex and varied combination patterns were commonly used for windows, while simple straight lattice and grid windows were no longer the mainstream choice. Simple straight bar windows also existed at this time, but they were different from the straight lattice windows of the Tang and Song dynasties, mainly due to the need to save labor and materials.

Straight lattice windows. Straight lattice windows are actually one of the earliest window styles to appear. Cross lattice windows were found on bronze artifacts from the Western Zhou Dynasty. In Han and Ming artifacts and paintings, various window forms can be seen, such as straight lattice, horizontal lattice, diagonal lattice, and lock pattern lattice, all of which evolved from the straight lattice window. The cross-section of a straight lattice window can be either square or triangular. The triangular straight lattice window is referred to as *pozi* lattice window in the *Yingzao Fashi*. The *pozi* window has its outward-facing straight bars vertically mounted in a rectangular window frame. The outward-facing slats allow light to enter, while the flat back surface is convenient for attaching window paper or screens. The spacing between the lattice bars is usually around one *cun,* and the number of lattice bars is often odd. Another type of straight lattice window with a square cross-section is called the *banling* lattice window. The *banling* lattice window is an early form of the straight lattice window, and the *pozi* lattice window developed from the *banling* lattice window.

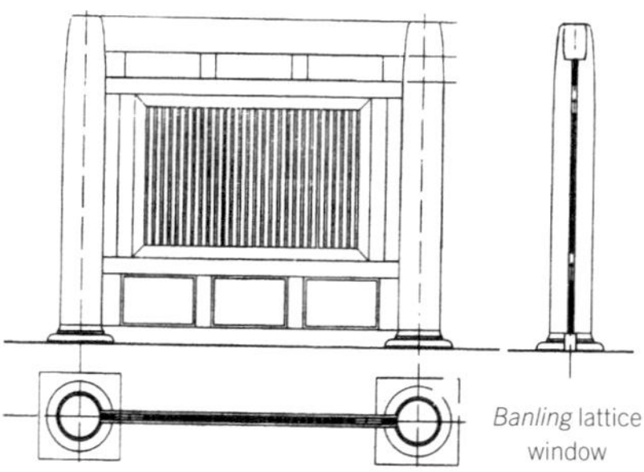

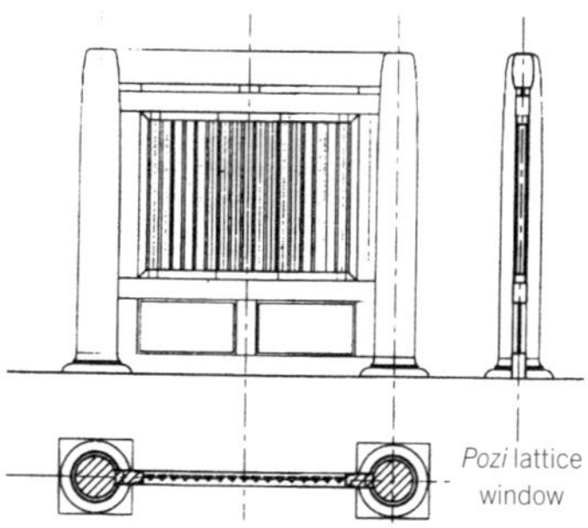

LEFT *Banling* lattice window.

RIGHT *Pozi* lattice window.

LEFT *Pozi* lattice window.

RIGHT Straight lattice window "one code and three arrows."

If the mullions of a straight lattice window are too long, a supporting root string can be added in the middle. In the early period, mortise holes were used for the supporting root string, and the lattice bars were connected by passing through the mortise holes. In the later period, a half-lap joint was made on both the supporting root string and the lattice bars. In the Qing Dynasty, three supporting root strings were placed on the upper, middle, and lower parts of the lattice bars, which was called "one code and three arrows." The straight lattice window can also be made with curved lattice bars. The *Yingzao Fashi* records the "*shandian* (lightning) window" and the "*shuiwen* (wave) window." These windows are typically two to three *chi* high and about two-thirds the width of the interval space. They are often used on the upper part of palace walls or on high mountain walls. The root strings of the lightning window are curved like water waves or lightning, hence the name *shandian* window.

Sill wall windows. The use of sill wall windows can be traced back to the late Tang Dynasty. They are called sill wall windows because they are installed on the threshold walls on both sides of a door or gate. They evolved from partition gates and are also known as partition windows. In terms of construction and shape, they are identical to the upper part of partition gates. The main difference between the two is that sill wall windows do not have the lower *zhangshuiban* and lower window tools found on partition gates.

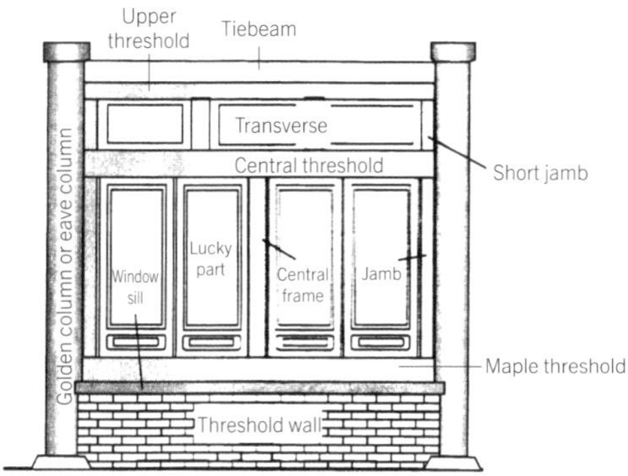

Front of sill wall window

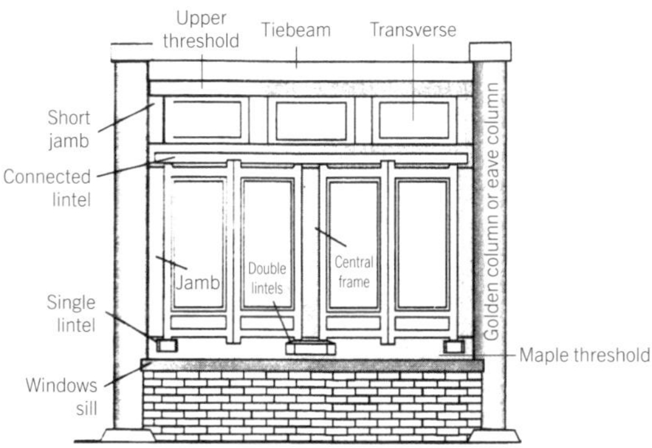

Back of sill wall window

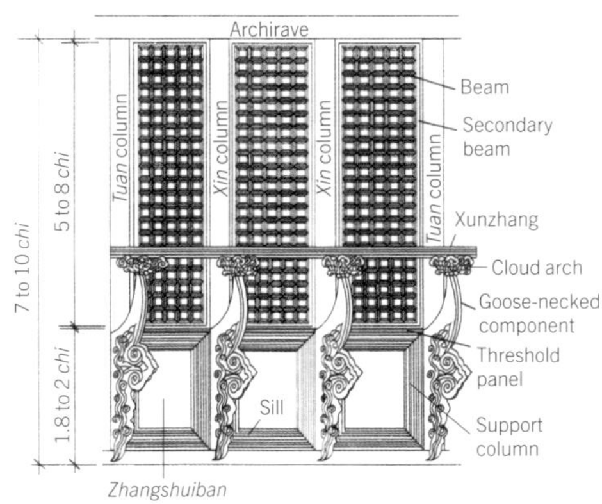

TOP Sill wall window.

BOTTOM Railing sill window in the *Yingzao Fashi.*

Removable windows. Removable window, also known as the hinged-latch window, is a window that can be supported or detached. It consists of two sashes, with the upper sash being able to be pushed out and supported (sometimes also pushed in), while the lower sash can be detached. The combination of the upper and lower sashes can be used as needed. Having only the upper sash supported makes the window lighter and more convenient, providing ventilation and rain protection and offering a certain degree of privacy to the interior space. In hot summers, removing the lower sash enhances ventilation. The proportions of the upper and lower parts of the removable window are generally in the ratio of 2:1 or 3:1. Removable windows can be traced back to the Han Dynasty and are commonly seen in Qing Dynasty architecture, where they were used extensively on threshold walls. They are most commonly used in residential buildings. They were also extensively used in palaces, temples, and gardens.

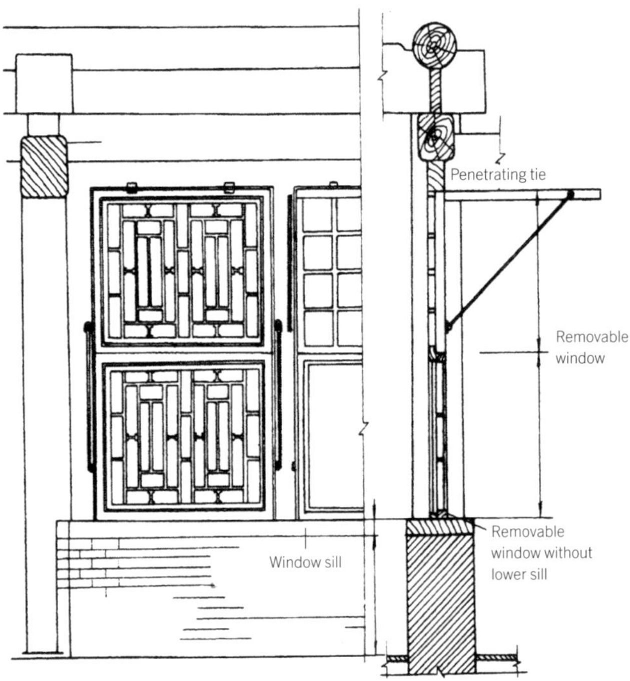

Removable window.

Removable window.

Transverse windows. Gates and windows should not be too large or too high, as this would be very inconvenient. In some taller buildings, there are transverse windows fixed above the gates or window frames. Transverse windows became widely used after the Song and Yuan dynasties, and their shape mainly follows the changes in the shape of the lower window.

In traditional Chinese timber structures, the shape of the frame structure allows for a variety of window types. In pavilions, corridors, walls, and gardens, the forms of windows are even more diverse. There are various types of window openings. The "leaking windows" with rich lattice patterns are known as "flower windows." These "leaking windows" can either allow some light to pass through or be completely open, and there are variations where only the patterns are visible without being transparent, known as "inlaid flower windows." Some windows have openings without actual window frames, and their shapes vary, such as the "hole windows," also known as "empty windows" or "assorted windows." In some walls, various forms of wall openings are left, creating a "flower wall." These "hole windows" can also be pasted with paper on both sides and equipped with lights, known as "light windows." Whether it's leaking windows or hole windows, they meet the need for ventilation and lighting while also satisfying aesthetic needs and often provide beautiful views. The shape of windows is actually very diverse and has gradually transitioned to brick and stone materials in later use. Common shapes include fan, hexagonal, square, bottle, gourd, peach, circular, and cross windows.

Partitions

Since the inception of architecture, its undeniable purpose has been to provide mankind with a secure haven, shielding them from the relentless forces of wind, scorching sun, harsh winters, and the lurking presence of creatures and insects. Thus, the existence of dwellings capable of protecting us from the outside world fulfills these essential needs. The notion of rational functional zoning and aesthetic considerations within the confines of a dwelling, however, emerges as a secondary or ancillary requirement once the primary objective of safeguarding against external threats is achieved. Even this secondary "requisite" has evolved over a long period of time. Should the need for new functional spaces or the desire for increased comfort in the home arise, the linear logic of thought would naturally lead to the creation of another "shelter from all directions" structure. Therefore, in the early stages of construction, the concept of interior zoning was not yet conceived.

At the same time, traditional Chinese wooden architecture, with its framework structure, encloses and roofs around the pillars to create a box-like space. In the construction process of these buildings, there is no need to consider interior space partitioning or division. The structure itself does not require internal walls for support. Adding internal walls would not only incur additional labor and material costs but also have no impact on fulfilling the need for "shelter from all directions." Even today, in some remote areas, self-built houses known as "three-room tile houses" are essentially "empty shell houses." They consist of a large open space inside; partition walls may be added inside depending on their function and if the owner has accumulated some financial resources; otherwise, the owners use furniture, curtains, or other methods to divide the space. Therefore, interior partitioning is not completed during the construction of the house. It deviates from the technical and developmental path of the building itself from the beginning. The formation of partitions is a purposeful exploration and accumulation driven by the continuous improvement of usage experience and the demands of real life. The framework structure of traditional Chinese wooden architecture provides great flexibility in terms of materials and structure for partitions, allowing for various possibilities and creating diverse forms. As a result, partition walls have become a distinctive feature of interior space treatment in ancient Chinese architecture.

In ancient Chinese literature and visual materials, there are numerous depictions of early forms of partitions such as curtains, screens, and folding screens. In the Han Dynasty, Sima Xiangru wrote in his poem "Changmen Fu," "The wind blows fiercely, ruffling the curtains and the canopy." The term "curtains" here refers to the tents used in military camps, which were then introduced indoors as "curtains" because of their simplicity and convenience. As a screen, it is recorded in *The Rites of Zhou*, "*Di*, the rear panel, refers to the use of a screen made of dyed feathers resembling the colors of phoenix." As a rear panel wall, the screen serves both as a wind barrier and a partition to divide the space. During the Qin and Han dynasties, there was a wealth of evidence showing that curtains and screens became essential fixtures in indoor spaces, forming a part of interior design. It is evident that the use of curtains and screens played a significant role in the development of interior partitions in ancient Chinese wooden architecture. Many terms related to curtains and screens can still be found in the *Yingzao Fashi*, such as screen doors (*pingmen*), curtain walls (*bizhang*), and decorative wall panels (*banzhang*).

There are approximately three types of interior partitions in timber constructions.

Closed full partitions. Full partitions use partition walls, which are closed walls that divide interior spaces. Depending on specific needs, materials such as soil, brick, and wood are used to build walls, creating various styles. The wall surface treatment can be modified as needed. The *Yingzao Fashi* records different forms of complete partitions, such as "partition boards or walls" (*jiejianbanzhang*), "compartment wall panels" (*xiangweiban*), and "decorative wall panels" (*bizhang*).

Functional half partitions. These partitions can take the form of exterior eaves or "interior lattice partitions in the palace hall," which are similar to partition gates. The difference lies in their position and function. They can also be integrated with interior furniture, such as antique shelves, bookshelves, storage cabinets, and screens. These partitions serve decorative, lighting, and ventilation purposes, preventing the interior space from being too closed and adding depth to the space. Another type of functional half partition is the *Taishi* (grand preceptor) wall, which is typically used in the back half of living rooms. It consists of a closed panel wall in the middle and passages on both sides, with the front of the panel wall being the focal point of the living room layout.

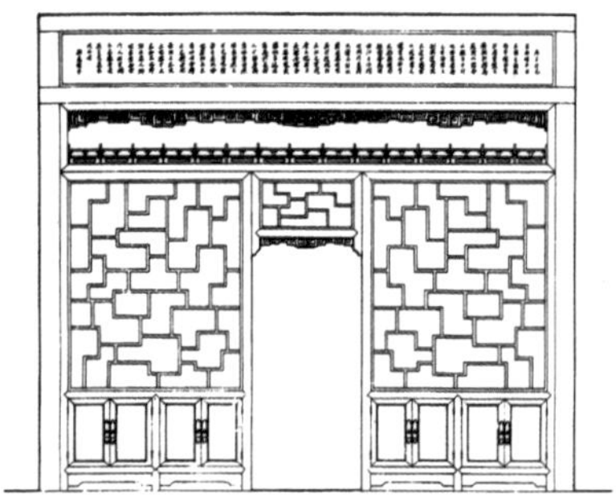

LEFT Antique shelf.

RIGHT *Taishi* wall.

Visual partitions. Visual partitions refer to the unique forms of partitions found in ancient architecture. Unlike the wall that separates the space, these partitions have a certain level of transparency or openness, allowing for a visual connection and a sense of openness between the divided spaces. The concept of visual partitions originated from "ornamental brackets" (*huayazi*), and they were commonly used in the Ming and Qing dynasties. There is no record of them before the Ming and Qing dynasties in the *Yingzao Fashi*. Visual partitions can be further classified into *tianwanzhao* (arched canopy), *luodizhao* (floor-to-ceiling partition), *langanzhao* (railing partition), and *huazhao* (flower canopy). There are also *chuangzhao* (bed canopy) used for smaller spaces.

Tianwanzhao, also known as *jiteluozhao* or *feizhao*, is attached to the top beams without touching the floor, and the sides are attached to the walls or columns. It is the simplest form of *huazhao* used to divide less prominent interior spaces.

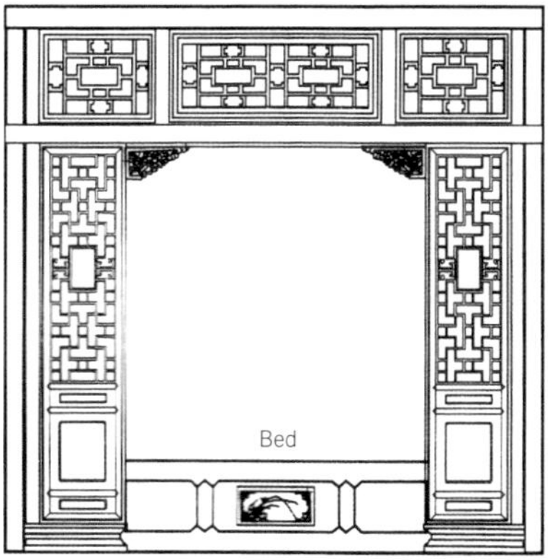

LEFT *Chuangzhao.*

RIGHT *Tianwanzhao.*

Luodizhao refers to the canopy corners on both sides of a *huazhao* that directly touch the ground. There are two forms: one extends the corners of the canopy to create a floor-to-ceiling style resting on wooden stools; the other forms partitions with the corners of the canopy and connects them to the wooden stools.

Langanzhao is mainly used to divide large interior spaces in houses. The entire canopy frame is divided into three openings, with a passageway in the middle and decorative partitions on both sides.

Huazhao is a combination of *tianwanzhao* and *langanzhao*. It has less transparency but stronger decorative features. Some are designed in trellis patterns, combining practicality with aesthetics. The shape of the central passage in a *huazhao* can vary, and there are variations called *huapingzhao* and *yueguangzhao*, depending on the contour style.

Whether it is *tianwanzhao*, *langanzhao*, or *luodizhao*, they can be made with carved wooden panels or wooden lattices. Generally, high-quality wood is selected for carving, leaving a margin for connection with the frame. The patterns and motifs are diverse, including continuous patterns arranged according to certain rules, such as palace flowers, diamond patterns, peonies, ice flakes, and plum blossoms, and irregular patterns, such as intertwining branches and vines.

Zhao in the Ming and Qing Dynasty architecture, plays an important role in the visual division of space. It later became an aesthetic standard for interiors, with decoration being more important than division.

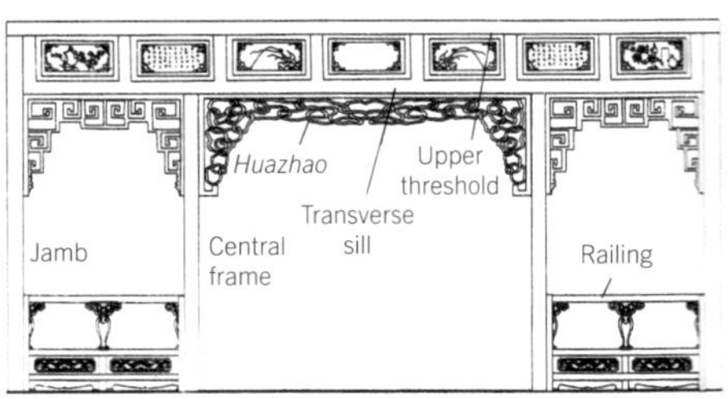

LEFT Partition *Luodizhao*.

RIGHT *Langanzhao*.

Suspended and Coffered Ceilings

In early wooden architecture, the internal roof structure was left completely exposed without any form of covering or treatment. This approach, known as *cheshang mingzao* in the Song Dynasty, differed from the direct exposure of the early wooden structures. *Cheshang mingzao* refers to a method of not treating the roof with any form of the ceiling, as opposed to the alternative approach of adding a ceiling. In the early stages of wooden architecture, there was little emphasis on the treatment of the ceiling, resulting in a complete lack of treatment. Therefore, *cheshang mingzao* in the Song Dynasty was a product of architectural technology and the level of comfort in housing, which differed from the lack of treatment in early ceilings on a conceptual level.

With the accumulation of experience and the gradual improvement of comfort awareness, people found that dust easily accumulated on some wooden frames, making them inconvenient to clean. The dust would often fall down when it accumulated to a certain extent. Moreover, the upper space was too large and had no direct spatial value. In addition, the climatic conditions in the northern regions had an impact on insulation and heating. These factors gradually led to the development of ceiling treatments. The treatment of ceilings was not limited by the roof and became more flexible, resulting in various forms. Spatial decoration became more important.

The aforementioned *cheshang mingzao* can be considered one of the intentional ceiling treatments. This type of treatment, also known as *mingfu*, involves exposing the roof structure and, whenever possible, refining the roof structure to enhance both its functional and aesthetic aspects. Due to the exposed structure, houses with this treatment have better ventilation and moisture resistance, especially in the rainy and humid regions of the south. The *cheshang mingzao* is commonly used in the southern regions and is sometimes referred to as *caofu* in local practices, which is essentially the same treatment. Apart from the intentional lack of treatment in the *cheshang mingzao*, ceiling treatments have developed into various forms, which can be broadly categorized into three types: suspended ceilings, coffered ceilings, and rolled ceilings.

Suspended ceilings. Suspended ceilings, also known as *chengchen*, have different forms and names in different periods. In the Song Dynasty, there were two different forms of suspended ceilings called *ping'an* and *pingqi*.

Ping'an is constructed by intersecting square timbers to form small grids, which are then covered with panels. The grids are usually two to three times the size of the square timbers, referred to as "one timber with two empty spaces" or "one timber with three empty spaces." "An" (闇) is a variant character of the word "dark" (暗), indicating that the upper wooden structure of the suspended ceiling is hidden and not visible. The overall shape of the *ping'an* resembles a square basin, and the connecting surfaces with the surrounding area form a sloping surface called *junjiao*. Examples of existing wooden structures using this technique include the East Hall of Foguang Temple in Wutai Mountain, Shanxi, and the Guanyin Pavilion of Dule Temple in Ji County, Tianjin. However, this form of suspended ceiling has been rarely used since the Liao and Song dynasties.

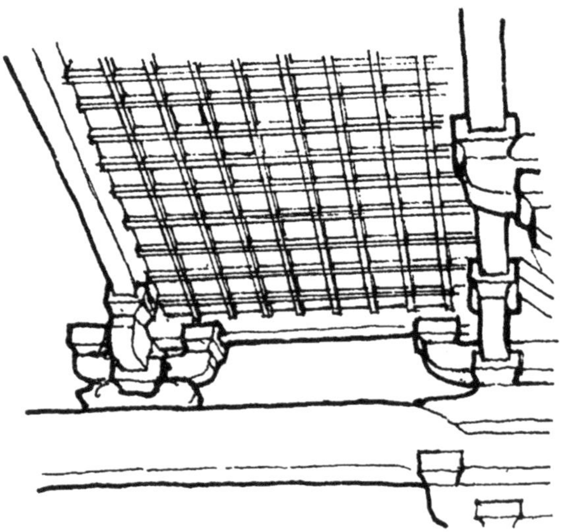

Ping'an in the Foguang Temple, Wutai Mountain, Shanxi Province.

TOP *Ping'an* in the Foguang Temple, Wutai Mountain, Shanxi Province.

BOTTOM *Ping'an* in the Guanyin Pavilion, Dule Temple, Ji County, Tianjin.

Octagonal well and *Piang'an* in the Guanyin Pavilion, Dule Temple, Ji County, Tianjin.

Pingqi is a form of suspended ceiling constructed by intersecting square timbers to form large grids. It is roughly "one timber with six empty spaces." In the Song Dynasty, the grids of the *pingqi* were adorned with intricate patterns called *luohua*, resembling a chessboard when viewed from above, hence the name *pingqi*. The square grids in the Song Dynasty were square, rectangular, and even polygonal, with relatively dense pattern designs inside. In the Ming and Qing dynasties, the grids were all square, with relatively sparse patterns. At the intersections of the timber frames, four inward-facing auspicious motifs were added to complement the patterns inside the grids. The timber frames formed a cross shape, and in the Qing Dynasty, this shape was also referred to as *jingkou tianhua*. The early decorative motifs inside the square grids included aquatic plants such as algae and lotus flowers, symbolizing the desire to ward off fire. However, this symbolism gradually faded over time. In the Song Dynasty, it developed into thirteen different patterns, including *panqiu* (coiled ball), *douba* (8-shaped pattern), and *diesheng* (stacked victories).[1] In the Ming and Qing dynasties, the main palaces used dragon and phoenix patterns for *pingqi*, while secondary buildings used floral patterns, using a composition of "round light with branch angles."

Pingqi in the Qing Dynasty.

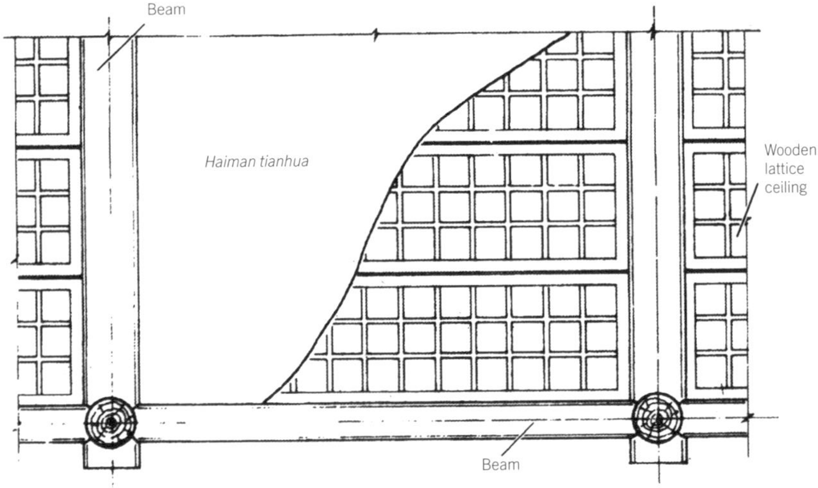

Haiman tianhua.

In addition to the well-known *jingkou tianhua* made of square timbers, the Ming and Qing dynasties also had suspended ceilings without using square timbers. This type involved attaching flat panels directly to the beams and then decorating them with patterns, which is known as *haiman tianhua* (wave-shaped ceiling).

Coffered ceilings. The coffered ceiling is a kind of special roof treatment system used in Chinese timber architecture. According to Li Yunhe, the coffered ceiling originated from a kind of "霤" (liú) structure and was retained as a roof form after the disappearance of the "霤" structure. According to Liu Zhiping's research, the "霤" structure is a roof framework that gradually narrows and is built layer by layer using angle beams on a square or polygonal plane without the need for a central column. It is an early form of roof structure.

Shen Kuo said in *Mengxi Bitan · Qiyong*, "The roof covered with lattices was called *qijing* or *zaojing* by the ancients, also known as *fuhai*. In today's language, it is called *douba*. The people of Wu call it *siding*. Only palaces, mansions, and temples are built with it." Even in the Tang Dynasty, there were clear restrictions on its use, stating that "houses below the rank of princes and nobles are not allowed to have heavy beam coffered ceilings." It can be seen that coffered ceilings were not used in ordinary houses but were reserved for higher-level palaces and temples, serving as a symbol of hierarchy.

From the perspective of its form, the appearance of the coffered ceiling can be traced back to an early period. In the tombs of the Han Dynasty, there were already examples of the *fudou* (covered bucket) and *dousi* (four-bucket) forms of coffered ceilings. Coffered ceilings were also found in stone caves from the Southern and Northern Dynasties, with lotus flowers and flying deities as decorative patterns on the walls of the coffered ceilings. In the Thousand Buddha Caves of Dunhuang, the *fudou* form was commonly used, with Buddhist stories and lotus flowers as the main decorative motifs on the inner walls and *taosi* or *dousi* form at the top. In addition, there were also examples of two *douba* coffered ceilings sandwiching one four-bucket coffered ceiling in the stone caves of the Southern and Northern Dynasties. The use of *dousi* or *douba* coffered ceilings continued in later periods after the Southern and Northern Dynasties.

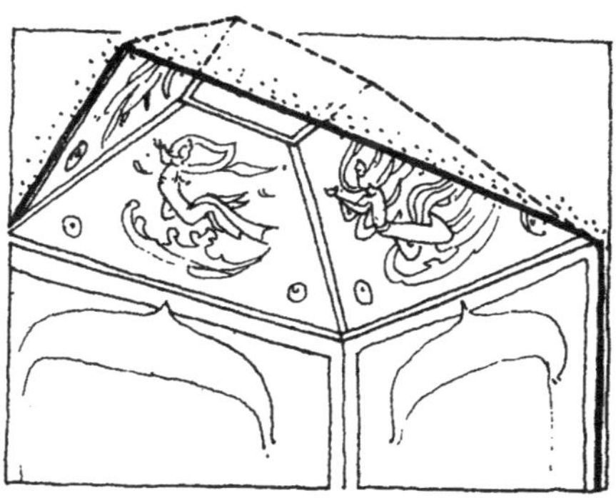

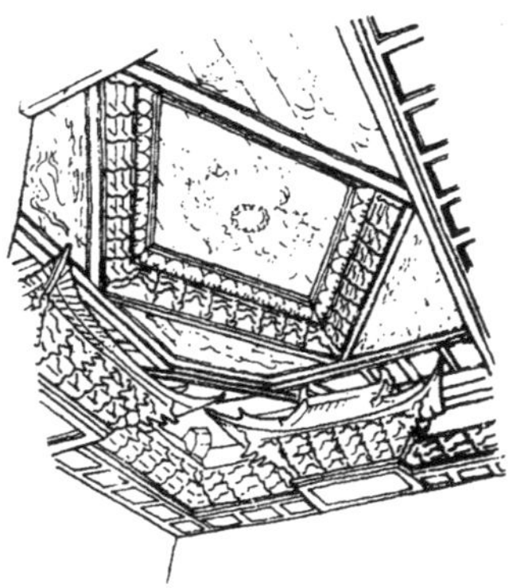

LEFT Suspended ceilings with the *fudou* form in Tianlongshan Grottoes, Taiyuan City, Shanxi Province.

RIGHT Diamond-shaped *fufou* well in the Main Hall of Jingtu Temple, Ying County, Shanxi Province.

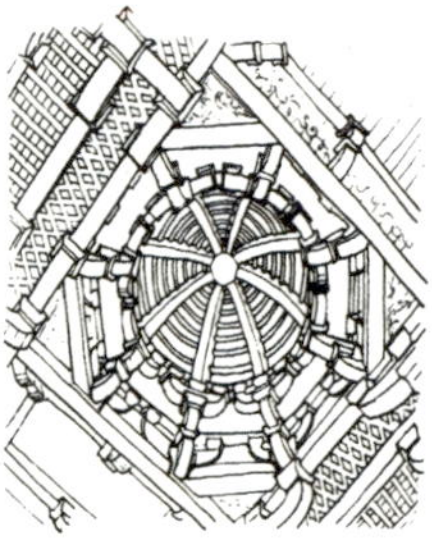

Perspective view

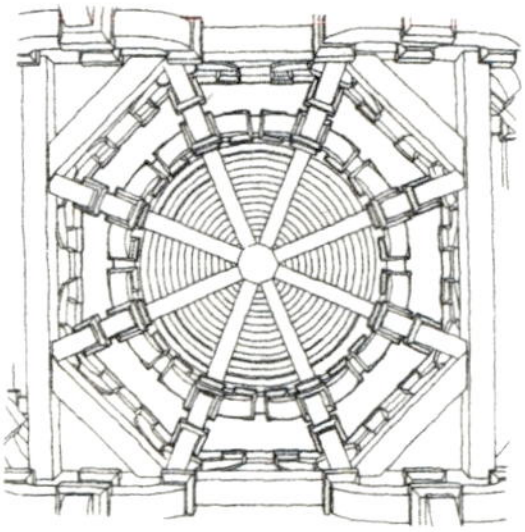

Bottom view

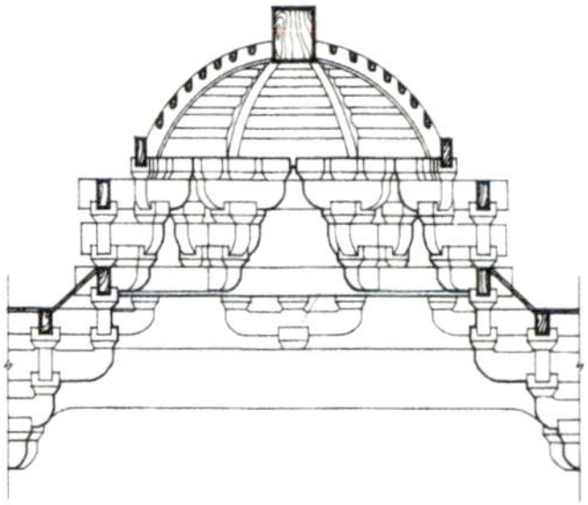

Elevation view

Round coffered ceiling in the Main Hall, Baoguo Temple, Ningbo City, Zhejiang Province.

During the Song Dynasty, the development of coffered ceilings reached a mature stage, with various forms and exquisite craftsmanship. The *douba* coffered ceiling was the most common form during this period. In the Song, Liao, and Jin dynasties, the use of *dougong* on coffered ceilings became more prevalent, resulting in more diverse and elaborate designs. Some coffered ceilings even featured small wooden carvings of heavenly palaces, symbolizing the abodes of immortals and Buddhas, on the inner walls. The tops of coffered ceilings in this period were either flat or domed. In the Ming and Qing dynasties, a new form of coffered ceiling emerged, known as the round coffered ceiling. The round coffered ceiling originated from the domed coffered ceiling, and it was often adorned with delicate bucket-shaped brackets, creating a highly decorative effect. However, some round coffered ceilings could appear too elaborate. In the later Ming Dynasty, the top of coffered ceilings began to feature larger *mingjing* (bright mirror), which symbolized skylights. In the Qing Dynasty, coffered ceilings with a dragon motif at the center of the top became popular and were also known as *longjing* (dragon well).

Coffered ceilings are mostly located in the main position of the hall space. In addition to the decorative function, they play an important role in emphasizing the space. The coffered ceilings in the Hall of Supreme Harmony in the Forbidden City were placed above the throne during the Ming Dynasty, while in temples, the coffered ceilings were positioned above the main hall and the statues of Buddha. This arrangement created a sense of prominence and spatial distinction for both emperors and Buddha statues.

Rolled ceilings. The rolled ceiling is a variant of the suspended ceiling form, mainly used in folk architecture or gardens in the south of the Yangtze River after the Ming Dynasty.

NOTES

1. Li Yunhe, *Principles of Classical Chinese Architectural Design* (Tianjin: Tianjin University Press, 2005), 28.

CHAPTER 8

THE ARTISTRY OF ARCHITECTURAL MURALS

The Emergence of Architectural Ornamentation

The emergence of architectural ornamentation can be traced back to the very beginning of conscious construction when buildings were intentionally designed for defense and spatial maintenance. In essence, architectural decoration has always existed, arising from conscious and functional choices. The main purpose is functional rather than aesthetic appeal. However, the existence of decoration is a conscious choice made by our ancestors, who already had an awareness of beauty. In other words, while the primary objective in architectural construction was functionality, they already had a level of consciousness regarding aesthetics.[1] Yang Hongxun stated, "The occurrence of architectural decoration is primarily aimed at beautifying the structural construction of architectural space. It is precisely because this decoration is combined with functional components that people can appreciate its beauty. Highlighting the characteristics and rational use of architectural materials, textures, and colors; emphasizing the mechanical characteristics and technicality of structural construction; and highlighting the functional characteristics and practical efficiency of components are the fundamental principles of architectural decoration."[2] Similar to our ancestors' conscious selection of spatial forms in architectural construction, this choice was deliberate and driven by specific functional goals. Moreover, the aesthetic sensibility possessed by our ancestors influenced the aesthetic and decorative outcomes of their choices. Likewise, in the selection of materials, components, and even the subsequent processing of materials, there was a conscious consideration for decoration.

This early sense of decoration was built upon the essentiality of materials and components, representing a conscious appreciation or deliberate choice for beauty after their utilization. The use of white plaster on interior floors and walls was discovered in the later period of the Yangshao culture and became widely employed during the Longshan culture. The use of white plaster during this period was not a necessary choice for the primary functional purpose of housing construction. Although white plaster also possessed functionality, its decorative nature had evidently become the primary consciousness of the initial users and was intentionally pursued. In other words, decoration transitioned from an initial conscious aesthetic choice to a deliberate stage. Patterns carved on white plaster walls were found in the Longshan culture site of Taosi Village, Xiangfen, Shanxi. In the Neolithic Hongshan culture site of Niuheliang, Liaoning, painted graphics and linear motifs were discovered adorning the walls of a temple interior. The paintings were created using red and white pigments on flattened and fired clay surfaces, depicting geometric patterns. The linear motifs involved creating raised flat lines or semicircular lines on the clay surface. These findings indicate that decoration had officially emerged as a purposeful presence in architecture. The use of decoration was also evident in contemporaneous discoveries of everyday utensils and tools, particularly in pottery and rock paintings. While the initial use of color and graphics may have been based on the functional needs of prehistoric religious forms such as totems and witchcraft, the strengthening of aesthetic consciousness gradually led to the formal use of purposeful decoration in architecture.

The origin of architectural decoration is rooted in functional choices, and even in its later development, it has always maintained a certain relationship with functionality. Decoration and functionality are rarely considered in isolation. The opposition to excessive luxury and waste has consistently influenced the development of ancient Chinese architecture. As a value orientation, it has been a manifestation of benevolent governance by rulers throughout history. Whether driven by a genuine pursuit of benevolent governance or compelled by traditional values and contemporary public opinion, rulers have always faced the dilemma of balancing frugality with grandeur and dignity in architectural construction. Both official and folk architecture have evolved under the influence of these two conflicting values. The opposition to excessive luxury and waste has always constrained architectural construction, with decoration being the first consideration in countering such excesses. Therefore, in ancient Chinese architecture, there was a prevalence of "decorative components," while "decorative structural elements" were relatively scarce. Even if they do exist, they must have a strong and legitimate functional justification, making it difficult to justify their presence solely based on aesthetics or artistic appeal. Decorative elements, such as the ridge-end ornaments resembling owls and the hanging fish and grass on the roof ridge, all carry symbolic meanings of blessings and safety. They serve as devices with functions of blessing and fire prevention, not merely as artistic aesthetic images.

The Colors of Architecture

Ancient Chinese architecture is characterized by vibrant and diverse colors, standing out among the world's architectural systems and representing one of the most richly colored architectural traditions. This is not only a reflection of aesthetic emotions towards color and social culture but also a result of the materials used and the construction methods employed in ancient Chinese architecture.

The formation of architectural colors is the result of a complex interplay of various factors. It is influenced by the characteristics of materials, functionality, social aspects, religious beliefs, spirituality, customs, rituals, symbolism, and regional influences. The specific use of colors on architectural components is simply the outcome of one dominant factor determining the choice of colors.

The basic colors of architecture are derived from the inherent qualities of the materials used, be it earth, stone, brick, wood, or others. These materials possess their own natural hues, which form the basis of the architectural color palette. As architecture evolves and adapts to social and spiritual changes, the transformation of foundational colors is influenced by cognitive factors. This color variation differs across different periods and types of architecture, reflecting the shifting societal values. Symbolism and ceremonial significance have always played a significant role in the use of color in architecture.

The establishment of the concept of color in ancient China undoubtedly underwent a process of development from simplicity to complexity. According to Chen Yanqing, the growth and evolution of color concepts should follow a natural and purposeful construction process. He inferred the growth of color concepts as follows: "1. Monochromatic—2. Initial differentiation into two colors (*yin-yang*, black-white, pure-mixed)—3. Three-color perception (black, white, red)—4. Four-color perception (black, white, red, yellow)—5. Five-color perception (black, white, red, yellow, green)—6. The five-color system under the dominance of mysterious color (mysterious yellow—black, white, red, yellow, green) and the emergence of intermediate colors."[3] From the available literature, we can see that during the Qin and Han dynasties, ancient China had already formed a clear color system. The concept of the five colors derived from the theory of *yin* and *yang* and the Five Elements constituted the fundamental five primary colors, from where intermediate colors were formed. It was recognized from the beginning that intermediate colors were considered secondary to the primary colors.

Throughout history, it has been perceived that the Chinese people possess a penchant for vibrant and extravagant colors, favoring hues of bold red and lush green. However, in the inception of the establishment of color concepts, pure colors (*zhengse*) were deemed superior. As the notion of colors took shape, they also acquired symbolic significance. According to the theory of *yin* and *yang* and the *Five Elements*, it is believed that the universe and all things in it are composed of five fundamental elements: water, fire, earth, metal, and wood. These elements interact with each other in a cycle of generation and restraint, creating a perpetual harmony. Wood gives birth to fire, fire gives birth to earth, earth gives birth to metal, metal gives birth to water, and water gives birth to wood. Every aspect of the natural world, including seasons, directions, and colors, is intricately connected and aligned with the principles of the Five Elements. The Five Elements correspond to the colors of "green, red, yellow, white, and black."[4] Green represents the gentle arrival of spring, resembling the budding leaves of wood. Its direction is the east, where the sun rises. Red symbolizes the scorching heat of summer, embodying the fiery essence. Its direction is the south, at high noon. Yellow signifies the element of earth, reflecting its earthy hue. Its position is at the center. White represents the refreshing coolness of autumn, mirroring the luster of metal. Its direction is the west, where the sun sets. Black signifies the chilling coldness of winter, embodying the essence of water and the depths of the abyss. Its direction is the north, in the midst of midnight. Each of the five colors holds its own unique symbolic meaning. When it comes to the use of colors in architecture, a clear understanding of their intended purpose can only be achieved through the lens of the Five Elements. Under the influence of the Five Elements theory, which has greatly influenced the later development of *feng shui* and other related disciplines, including ceremonial practices, the selection of colors in architecture has become imbued with profound symbolism and ceremonial significance.

In *The Book of Rites*, it is stated, "For the imperial palace, the pillars are vermilion; for the feudal lords, they are dark gray; for the high-ranking officials, they are pale blue; and for the commoners, they are dark blue." During the

Zhou Dynasty, the pillars and walls of the imperial palace were painted red, a tradition that continued in the palaces and government offices of the Han Dynasty. Red became the designated color for individuals of significant status. Although yellow, representing the earth and symbolizing power, later became the highest color, red still retained its sense of nobility. In important buildings such as palaces and temples, walls were painted vermilion, and doors were known as *zhumen* (vermilion doors) because they were adorned with vermilion lacquer.

Throughout the dynasties, there has been a significant emphasis on the hierarchical color system for architectural components. During the Ming Dynasty, specific regulations were established for the main gates of different ranks. For example, the main gate of a prince's residence was adorned with vermilion lacquer, golden nails, and copper rings, while the main gate of the high-ranking official's residence featured green oil paint and copper rings. The officials of the middle rank, such as the dukes and marquises, have their doors adorned with golden lacquer and tin rings in the shape of animal faces. For officials of the first and second ranks, the doors are adorned with green lacquer and tin rings in the shape of animal faces. For officials of the third to fifth ranks, the doors have black lacquer and tin rings. As for officials of the sixth to ninth ranks, their doors are adorned with black lacquer and iron rings. In the Qing Dynasty, yellow-glazed tiles were reserved exclusively for palaces and temples, while the residences of nobles were only allowed to use green-glazed tiles. Ordinary people were limited to using gray cloth tiles. The decorative colors for buildings were also subject to hierarchical regulations. In general, the color hierarchy for buildings was as follows: yellow was the most esteemed, followed by red, green, azure, blue, black, and gray. Palaces used gold, yellow, and red tones, while ordinary residences were limited to white, black, and gray for wall and roof colors.

Whitewashing and Painting

In the early stages of human society, people gradually became aware of pigments derived from minerals and plants, which were then applied to daily life, tools, and building materials for both protection and decoration purposes. The use of paint, in particular, not only served the functional purpose of protecting wood from moisture but also allowed for a wider range of colors. In the early stages of the Yangshao culture, evidence of applying a thin layer of mud to the bottom or walls of dwellings has been found. Similarly, in the Banpo site, a smooth white mud coating was discovered, indicating that early architectural maintenance practices involved the application of mud to achieve a smoother and more sealed surface. As the practice developed, red soil, white lime, and powdered shells gradually came into use, making the process of plastering smoother and more colorful. At least during the Shang Dynasty, *shenhui* (shell ash) was already being applied to mud walls.[5] In excavated architectural components from the Yanshi Shangcheng site, a clear white lime coating can be seen, with a smooth layer of fine mud at the base and a white lime coating on the surface, indicating a certain level of craftsmanship. During the Zhou Dynasty, there was a practice of painting walls white and floors black. In palaces, pillars, walls, and other elements were painted red, and specialized painting tools were already in use. By the Qin Dynasty, the floors of the Xianyang Palace were already painted red, and during the Han Dynasty, palaces had walls, pillars, and foundations painted red. Descriptions of *danying* (vermilion pillars) and *zhuque* (vermilion gates) can be found in historical texts, indicating the tradition of painting the exterior walls of palaces, government offices, and temples red, which continued into the Qing Dynasty. Even in modern architecture, the practice of plastering walls is still widely used, primarily for aesthetic purposes.

From the archaeological discoveries of lacquered wooden objects, it can be seen that the use of lacquer was already mastered during the late Shang and Western Zhou periods. Lacquered wooden coffins have been found in tombs from the Warring States and Western Han periods, and traces of painted walls have been discovered in some burial chambers. The lacquer craft underwent significant development, allowing for the widespread use of lacquerware and the formation of unique artistic categories. Lacquer also played an important role in ancient Chinese architecture, providing essential moisture and waterproofing protection for wooden structural materials. Due to the unique characteristics of wooden components in Chinese architecture, lacquer became one of the most important decorative materials, even the most crucial one for architectural coloring. It contributed to the splendid and colorful style of ancient Chinese architecture and formed a distinctive architectural painting system that sets it apart from other architectural systems.

Color Painting

From the use of color on everyday objects, especially colored pottery, in primitive societies, we can infer that our ancestors had the ability to use color and draw graphic designs as decoration. However, it is difficult to determine when this ability was applied to architectural decoration and the emergence of architectural painting due to the lack of clear physical evidence. From the existing literature and pictorial materials, including the tombs of the Han Dynasty, we know that architectural painting existed in some form on the beams during that period. But what about before the Han Dynasty? When did it start and when did it become more established? In the Neolithic Hongshan culture site of Niuheliang Goddess Temple in Liaoning, geometric patterns were painted on the compressed and fired clay walls using red and white pigments. Although they were not directly applied to wooden structures, they can be considered an early formative stage in the development of architectural mural painting.

From the large number of exquisite lacquer ware unearthed from the Spring and Autumn and Warring States periods, it is evident that the use of lacquer to protect and decorate wooden components was already mastered during that time. In the Eastern Zhou, Warring States, and Qin dynasties, metal components were widely used in palace architecture to reinforce and decorate wooden components at junctions. These components were called "jingang." However, *jingang* gradually disappeared during the Han Dynasty, and there is no evidence of their existence in the Tang and Song Dynasties. With the development of wooden structures, *jingang* (decorative beam end-cap) lost its functional role in reinforcement, and the extensive use of architectural painting emerged following the disappearance of *jingang*. Architectural painting imitated the patterns of *jingang* in the original locations, clearly adopting the decorative motifs of *jingang*.[6] While it cannot be said that architectural painting emerged solely because of the disappearance of *jingang*, it is an undeniable fact that architectural painting flourished after the disappearance of *jingang*. Eventually, architectural painting developed into a mature form, with the *gutou* (end portion of the painted beam) and *zaotou* (an intermediate portion of painted beam) at the ends of the beams still retaining the characteristics of *jingang* decoration.

According to historical records, there were colorful paintings of aquatic plants on the palace ceilings during the Qin and Han dynasties. Not only were the ceilings decorated, but colorful brushwork was also used on pillars, lintels, and doors and windows, gradually giving rise to patterns such as brocade designs.

During the Wei, Jin, and Southern and Northern Dynasties, the common practice of adorning walls with red pillars and white walls gradually reduced the use of figurative decorations. Instead, the brackets and rafters often featured continuous patterns of intertwining vines and branches.[7] Influenced by the introduction of Buddhism, motifs such as lotus flowers, flames, and celestial be-

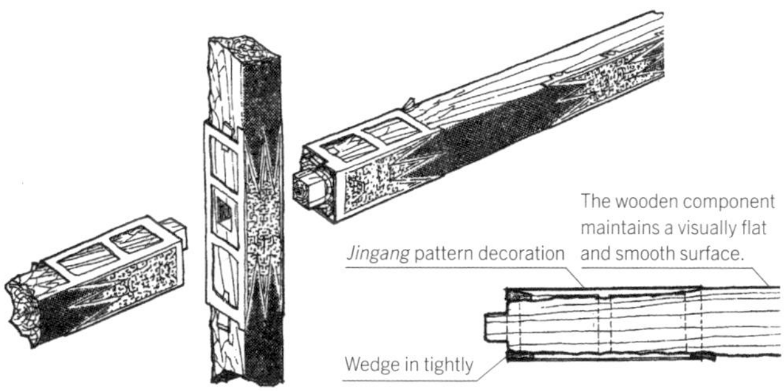

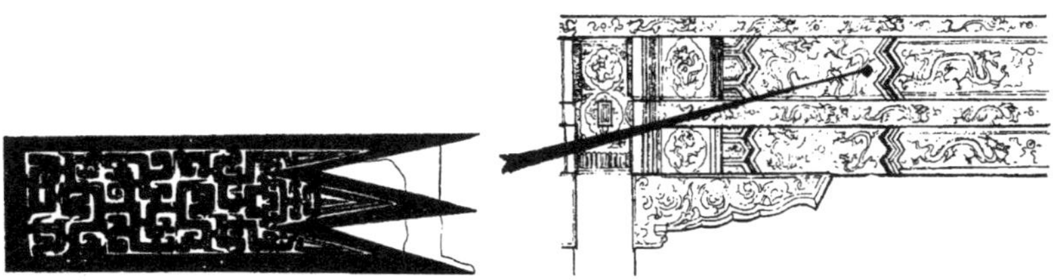

Hexi paintings in the Qing Dynasty, with the *gutou* and *zaotou* still retaining the shape of "*jingang* tooth."

TOP *Jingang* for buildings before the Han Dynasty.

BOTTOM "*Jingang* tooth" on the color paintings in the Qing Dynasty.

ings also found their place. It was during this period that the technique of *dieyun* emerged, wherein the same color was used to create gradual variations of light and shade by employing two to four different tonal levels.

During the Sui and Tang dynasties, the basic forms of decorative motifs inherited from the Southern and Northern Dynasties were largely preserved. The standardized usage of decorative patterns had also been established.

In the Song *Yingshi Fashi*, the art of color painting is categorized into six distinct styles: "multicolored decoration," "grinding jade attire," "green-blue layered haze between edges attire," "unraveling green attire," "vermilion powder brush decoration," and "mixed-style decoration." The book also provides detailed information on the process of color painting. During the Song dynasty, there was a transition from warm color tones to predominantly cool color tones, which had a significant influence on subsequent developments. Techniques such as graduated shading from dark to light and from light to dark were widely employed. The focal point of colored decorations shifted to the *lan'e*, forming a distinctive style different from previous dynasties. The composition of the *lan'e* clearly featured two decorative elements at the ends and *fangxin* (a central lintel), a style that would be followed in later periods.

During the Yuan Dynasty, a new decorative pattern known as *xuanzi* (spiral pattern) emerged, which had a significant influence on the Ming and Qing dynasties. This pattern was widely utilized during the Ming and Qing periods.

During the Ming dynasty, two major decorative styles, namely *jinlong* (golden dragon) and *xuanzi* paintings, were established. Unlike the Song and Qing dynasties, it was customary not to depict any images on the *fangxin*. In the Qing Dynasty, three major categories of color paintings emerged: *hexi* (imperial seal) color paintings, *xuanzi* color paintings, and Su-style color paintings.

Color painting in the Song Dynasty

In the Song Dynasty, color paintings were categorized into different grades based on their usage in various buildings. Different grades of architecture used different levels of architectural color paintings. Generally, the top-grade color paintings were the "multicolored decoration" and the "grinding jade attire," while the middle-grade ones were the "green-blue layered haze between edges attire" and the "unraveling green attire." The lower-grade color paintings included the "vermilion powder brush decoration" and others.

Multicolored decoration. Multicolored decoration refers to the use of five colors: green, yellow, red, white, and black, representing a rich color palette. Multicolored decoration is characterized by a green-blue layered haze as the outer edge, a red-colored ground in the center, and the application of various colorful patterns. Alternatively, it can feature a vermilion-layered haze as the outer edge, a green-colored ground in the center, and the same colorful patterns. This creates a vibrant and magnificent architectural color painting, making it the most colorful form of color painting. Multicolored decoration was the highest-grade specification in the Song Dynasty and was commonly used in important buildings such as palaces and temples. It was primarily applied to wooden components such as beams, brackets, columns, lintels, and rafters.

Multicolored decoration was used on beams and brackets, with colored lines drawn around the outer edges, known as *yuandao* (edge paths). The *yuandao* was adorned with green, blue, or vermilion layered haze, and within the *yuandao* frame, various colorful patterns were painted, with vermilion, green, or blue used for the ground. According to the *Yingzao Fashi*, the main floral patterns included pomegranate flowers, precious flower motifs, lotus flowers, round gem motifs, intertwined circles, leopard foot motifs, agate grounds, fish scale motifs, and persimmon motifs. There were six types of intricate patterns, including intricate motifs, mat patterns, turtle shell motifs, four corners, square rings, and meandering water motifs.

The multicolored decoration on columns is divided into three parts: the capital, the shaft, and the base. The capital is adorned with intricate or intricate motifs, while the shaft and the capital are harmoniously decorated with the same intricate motifs. The base is painted with green or red layered haze lotus flowers.

The eaves rafters are painted with green or red layered haze lotus flowers, or they can be adorned with flame motifs or bright pearls. The flying rafters are decorated with green or blue interlaced beads and layered haze along with angular lines. They can also be painted with square motifs or pointed motifs.

Grinding jade attire and green-blue layered haze between edges attire. These are two different decorative styles. Grinding jade attire features a green-

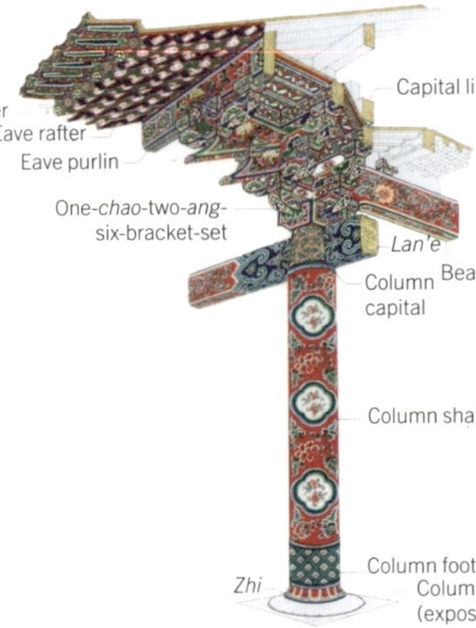

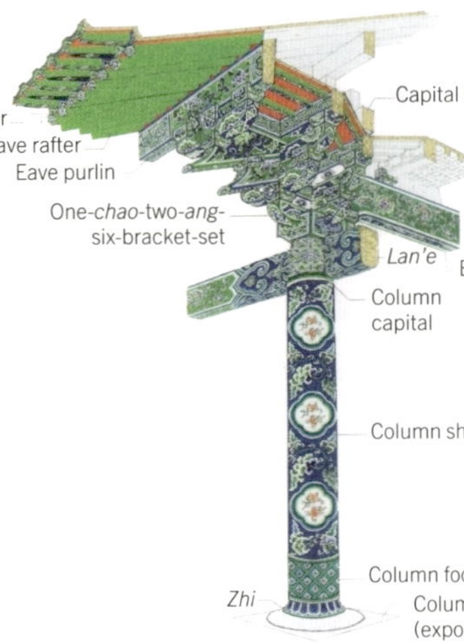

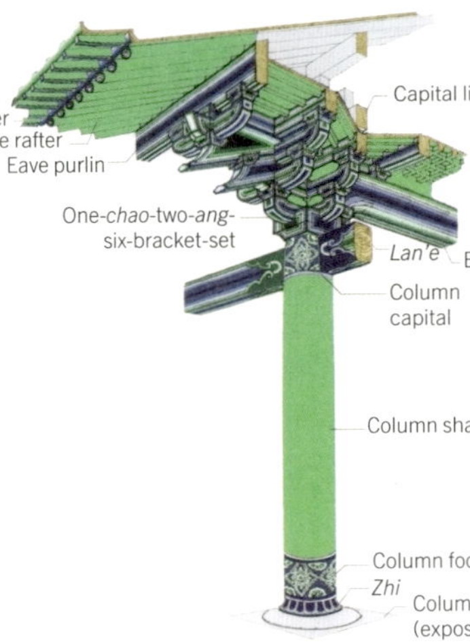

TOP LEFT Multicolored decoration.

MIDDLE RIGHT Grinding jade attire.

BOTTOM LEFT Green-blue layered haze between edges attire.

blue layered haze as the outer edge, with a light green floral pattern on a green background in the center. Green-blue layered haze between edges attire has the same outer edge as grinding jade attire, but the center does not feature any patterns, only the green-blue layered haze. These two styles, apart from multicolored decoration, are known for their exquisite craftsmanship and were widely used in residential buildings, gardens, and secondary palaces. They were introduced during the Song Dynasty and had a significant influence on the Ming and Qing dynasties.

The floral patterns of the grinding jade attire are described in the *Yingzao Fashi* as being the same as those of the multicolored decoration. The depiction of wooden components is also divided into two main parts: the beams and columns.

Unraveling green attire, unraveling green and adding decoration attire, and vermilion powder brush decoration. These forms of color painting all originate from the ancient red and white color paintings and primarily use warm colors such as red ochre. Unraveling green attire involves applying a coat of red ochre on the structure and adding a green-blue layered haze border on the outer edge. The difference between unraveling green and adding decoration attire and unraveling green attire is that the former includes decorative patterns painted on top of the red ochre base color. Vermilion powder brush decoration is characterized by a coat of red ochre on the surface and a white border on the outer edge. Yellow ochre brush decoration is a variation that replaces red ochre with yellow ochre. These painted decorations are considered to be of lower quality and are mainly used in secondary buildings. In recent years, the practice of "seven red and eight white," which refers to the use of red ochre and white in painting, a type of vermilion powder brush decoration, has been discovered in some surviving buildings and tombs from the Tang, Song, and Han dynasties.

Mixed-style decoration. Mixed-style decoration is a colorful painting technique that combines different types of painting in an intertwined manner, such as the combination of multicolored decoration and grinding jade attire, or the combination of green-blue layered haze between edges attire and grinding jade attire.

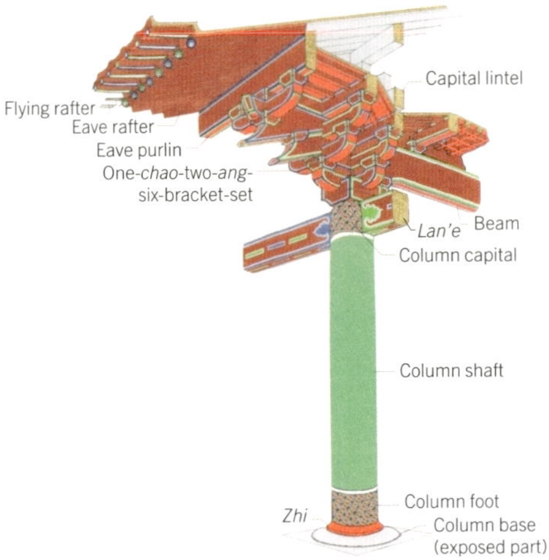

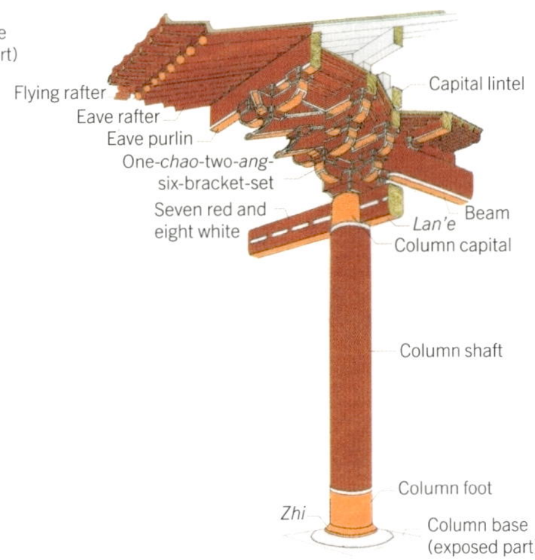

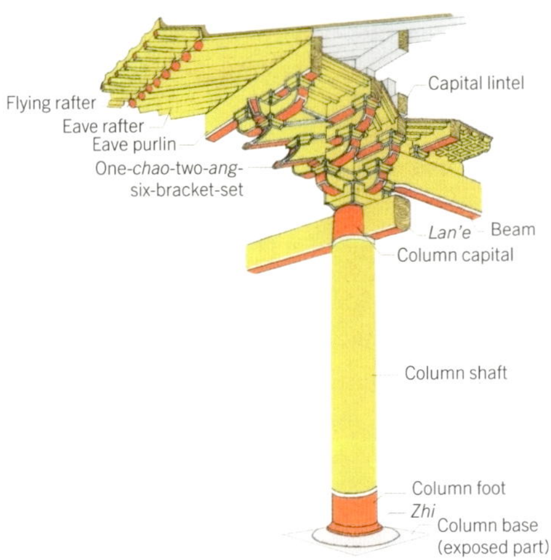

TOP LEFT Unraveling green attire.

MIDDLE RIGHT Vermilion powder brush decoration.

BOTTOM LEFT Yellow ochre brush decoration.

Color painting in the Qing Dynasty

Qing-style color paintings are divided into three categories: *hexi* color paintings, *xuanzi* color paintings, and Su-style color paintings. The composition of these three types of color paintings mainly consists of horizontal plaques, cushion boards, and door frames, forming a scroll-like picture with horizontal length. The picture is divided into three sections, also known as *fensanting* (three stops). The boundaries between the three sections are called *santingxian* (three-stop lines). The door frame is located within the three stops, with the *zaotou* (also known as *zhaotou*) on the left and right sides, and the *gutou* outside the *zaotou*. The patterns within the *gutou* are called *hezi* (boxes). The lengths of the *gutou*, *zaotou*, and door frame can be adjusted according to the horizontal length. Lines are used to separate the different sections, including the *fangxin* line, *gutou* line, *hezi* line, fork line, and strip line, also known as the "five major lines."

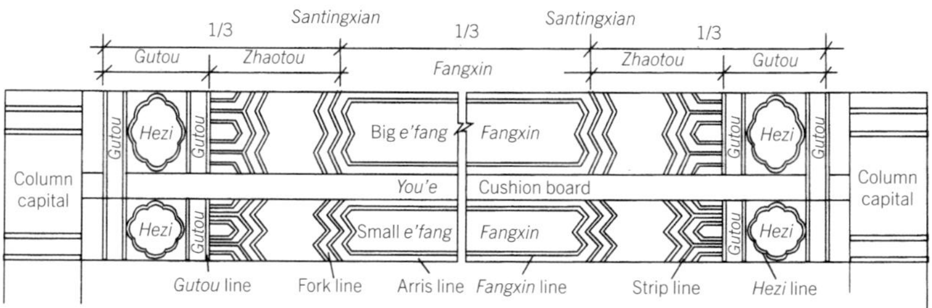

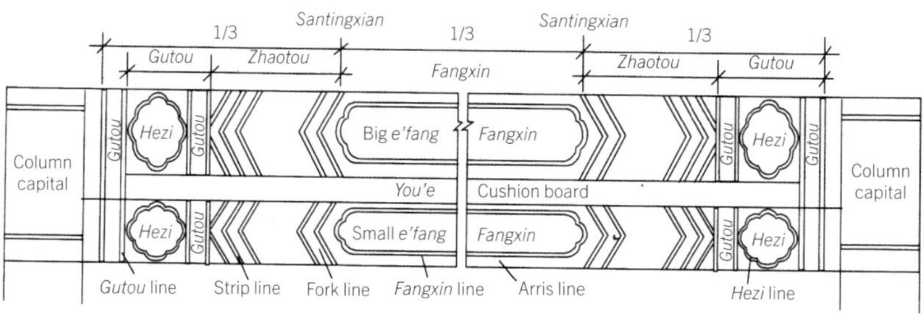

TOP Beam pattern of *hexi* color paintings.

BOTTOM Beam pattern of *xuanzi* color paintings.

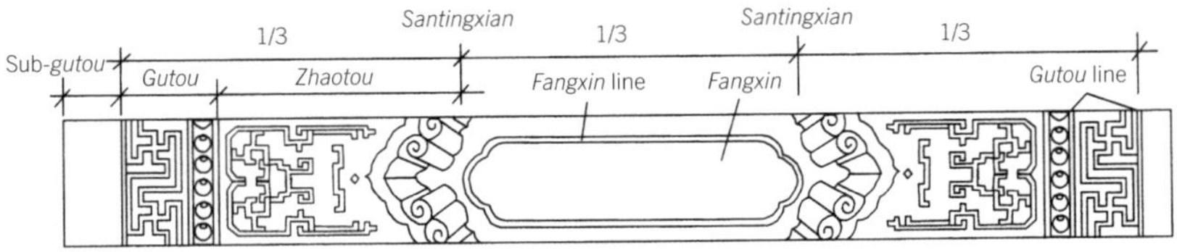

Beam pattern of Su-style color paintings.

***Hexi* color paintings.** *Hexi* color painting is the highest level of color painting in the Qing Dynasty and is mainly used in important halls of royal buildings. In *hexi* color painting, the boundaries between different parts, such as the *gutou*, *zaotou*, and *fangxin* are separated by folded lines instead of straight lines. The *fangxin* is connected to the column head with straight lines on both sides, creating a unified appearance in the colored decoration of the column frame. In the three-stop lines of *hexi* color painting, there are clear triple boundaries between the *fangxin* and *zaotou*, as well as between the *zaotou* and *hezi*. The colors used in *hexi* color painting mainly include green, blue, red, and purple. The large door frame and small door frame in the eaves are predominantly blue-green, while the middle part is adorned with bright red cushion boards to create a strong contrast of warm and cool colors. Sometimes, the technique of applying gold leaves to the main lines and patterns is also used in *hexi* color painting, with the number of gold leaves varying between large and small applications. White powder may also be added to the colors to control the brightness variations.

In *hexi* color paintings, dragons are often depicted in the *gutou*, *zaotou*, and *fangxin* sections. Other major patterns include phoenixes, auspicious grass, lotus flowers, and *lingzhi* mushrooms. Depending on the specific patterns used, *hexi* color paintings can be classified as *jinlong hexi* (golden dragon), *longfeng hexi* (dragon and phoenix), *longcao hexi* (dragon and grass), and so on. The color schemes in these paintings are generally consistent, with the level of intricacy and detail decreasing in each category.

Jinlong hexi is the highest level in *hexi* color paintings, characterized by the depiction of golden dragons. All the lines are adorned with gold leaf. The central design of the *e'fang* shows two dragons playing with a pearl, with a

Hexi color paintings.

Hexi color paintings.

background color of either green or alternating green and blue. For example, a green background is used for the big *e'fang*, while a blue background is used for the small *e'fang*, or they can be used interchangeably. The *zaotou* also has dragon patterns. For longer distances, ascending and descending dragons are used, while for shorter distances, ascending dragons are used on a green background, and descending dragons are used on a blue background, although they can be interchanged. The *hezi* design features a seated dragon, and the sides of the *hezi* are adorned with interlocking patterns. The central cushion board can feature various patterns of swimming dragons or alternating dragon and phoenix designs, while the wood-plate lintel uses horizontal dragons from either end towards the center.

Longfeng hexi is the second-highest level of *hexi* color paintings, following *jinlong hexi*. The main theme of the color painting is the depiction of dragons and phoenixes. The door frame's central design features auspicious patterns such as dragons and phoenixes bringing blessings and twin phoenixes symbolizing good fortune. The *zaotou* features alternating ascending and descending dragons, as well as flying phoenixes. The *hezi* pattern features seated dragons and flying phoenixes. The *gutou* has no interlocking patterns, and there is empty space between the *gutou* lines. The wood-plate lintel and the cushion board lintel feature one dragon and one phoenix.

Longcao hexi is one level lower than *longfeng hexi* in terms of hierarchy. The main theme of the color painting is the depiction of dragons and grass patterns. Different from *longfeng hexi*, the *longcao hexi* door frame's central design, *zaotou*, and *hezi* primarily consist of golden dragons and grass patterns. The big *e'fang* features two dragons playing with a pearl, while the small *e'fang* features auspicious grass patterns. The *zaotou* depicts ascending or descending dragons, and the *hezi* features lotus flowers. The *gutou* is depicted as a static pattern, and the wood-plate lintel and cushion board lintel have cogwheel grass patterns.

***Xuanzi* color paintings.** *Xuanzi* color painting is one level lower than *hexi* color painting. They feature striking patterns of petals with swirling and rotating motifs. They are mainly used in secondary buildings of palaces, *yamen* (government offices), temples, and other architectural structures, but are not generally used in ordinary residences. The overall composition of *xuanzi* paintings consists of the *gutou* and *zaotou* at each end, which together occupy one-third of the

Xuanzi color paintings.

total length, and the middle *fangxin*, which also occupies one-third. In other words, the composition is divided into three parts: two-thirds at the ends and one-third in the middle. The patterns inside the *gutou* are called *hezi*, which can be categorized as *zhenghezi* (whole boxes), *pohezi* (broken boxes), and *haitang hezi* (begonia boxes). *Zhenghezi* features a complete floral pattern inside the *gutou*, with each corner representing a quarter of a flower, forming a unified unit pattern. *Pohezi* divides the pattern inside the *gutou* into four parts with diagonal lines, with each part containing half of a floral pattern. *Haitang hezi* has a pattern inside the *gutou* in the shape of a begonia flower. The patterns inside the *zaotou* also have whole and broken variations. One complete swirling motif is considered a whole, while half of a swirling motif is considered a broken. The *zaotou* can have various combinations, such as one whole and two broken, or two whole and two broken. The *fangxin* can feature various patterns, including single-color brushwork, brocade patterns, grass patterns, and dragon patterns. In terms of overall style, *xuanzi* color painting can be either elaborate and colorful or simple and elegant, depending on the variation of colors used.

In *xuanzi* color paintings, the *zaotou* features a fixed spiral flower pattern. The spiral flower consists of a central pattern of the flower called the "spiral eye," the gardenia jasminoides petals, the triangular areas between the petals called *lingjiaodi*, the angled parts formed by the intersection of the spiral flower, and the outer shape called *baojiantou*. There is a clear double boundary line between the *fangxin* and the *zaotou*, as well as between the *zaotou* and the *gutou*. The *gutou* in *xuanzi* paintings has no pattern and is also known as "dead *gutou*," which is usually depicted with a green or blue background.

There are eight different types of *xuanzi* color painting, based on the amount of gold leaf used. These types include *jinzuo moshi nianyu* (a large amount of gold leaf is used, creating a rich and luxurious appearance); *yanzuo moshi nianyu* (a moderate amount of gold leaf is used, creating a smoky and elegant effect); *jinxian dajindian* (gold lines and large gold dots are used, creating a bold and eye-catching design); *moxian dajindian* (black lines and large gold dots are used, creating a contrasting and striking effect); *jinxian xiaojindian* (gold lines and small gold dots are used, creating a delicate and intricate pattern); *moxian xiaojindian* (black lines and small gold dots are used, adding subtle accents to the design); *yawumo* (a combination of elegant ink brushwork and subtle gold leaf accents); *xionghuangyu* (the use of yellow jade color, creating a unique and vibrant effect).[8]

Su-style color paintings. Su-style color painting is a form of folk art that originated in the Jiangsu and Zhejiang regions of China. In the early days, they were used by craftsmen in Suzhou for the palaces of the Southern Song Dynasty. Su-style color painting is known for its freedom, vibrancy, and relaxed style. They are closer to everyday life and are mainly used in lower-level architectural structures such as palaces, gardens, and other residential buildings. The tech-

Su-style color paintings.

Su-style color paintings.

niques used in Su-style painting are relatively free and flexible. The connecting line between the *fangxin* and *zaotou* can take various forms, such as curling grass, smoke and clouds, and angular lines. There are no strict rules or regulations. The *gutou*, *hezi*, and *fangxin* sections can all be used freely and flexibly, allowing for artistic expression and creativity.

Su-style color painting is divided into different forms, including the *fangxin* style, *baofu* style, *haiman* style, etc. Among them, the *baofu* style is particularly distinctive. It features a semicircular *baofu* (bundle) at the center, surrounded by a composition of beams, cushion boards, and *e'fang*. The *baofu* can be filled with various lively and freehand patterns, depicting landscapes, historical stories, pavilions, animals, and more. The techniques used in Su-style color painting are flexible and diverse, including pattern-based and painting-based approaches. The shape of the *baofu* also varies, allowing for various creative expressions. The *baofu* line serves as the boundary of the *baofu* composition and can take different forms, such as smoke and cloud fading frames, patterned borders, scroll borders, and curling grass borders. The smoke and cloud fading frame, also known as *tuiyunyan*, is created by layering several fading layers from the outer frame line inward. The smoke and cloud frame consists of a *tuozi* (support) and a *tong* (tube). The *tuozi* is a series of continuous folded lines, gradually becoming sparser towards the center and fading layer by layer, sometimes with as many as six or seven layers. The *tong* is a cylindrical arc line that fades layer by layer, with more than ten layers of fading.

The *haiman* style painting is characterized by even more freedom in its artistic techniques and flexible composition compared to the *baofu* style painting. In *haiman* style color painting, there are no boundaries for the *fangxin* and *zaotou*, allowing for integrated painting. The composition is not limited to the *gutou* and the patterns within the *gutou*, but can extend to the *e'fang*, cushion board, and frame. This allows for greater artistic expression and flexibility in the placement of various elements within the painting.

Color paintings on other building parts in the Qing Dynasty

Color paintings on *dougong*. In the Qing Dynasty, color painting on *dougong* consists of two parts: the body and the bracket board. The body part mainly involves coloring and drawing lines. The coloring refers to the base color of the body, which is usually a combination of green and blue, with the column head as the reference point. The *sheng* and *dou* components are typically painted with a green background, while the *gong*, *qiao*, and *ang* components are painted with a blue background, or vice versa. The *gongyan* is painted in red, and the colors are used interchangeably between the different components. Lines are drawn along the edges of the components, and there are two options: gold lines or colored lines. Colored lines are usually black or yellow, and the choice of which type of line to use depends on the overall color scheme of the painting.

The color paintings on the bracket board mainly feature patterns such as sitting dragons, dancing phoenixes, flaming treasure beads, and Bodhi grass. The choice of patterns is also based on the selection of the main painting of the wooden structure. If the main wooden structure painting is a higher-level and more formal type of color painting, the bracket board should have patterns such as sitting dragons, dancing phoenixes, Bodhi grass, or flaming treasure beads. For color painting below the level of *xuanzi*, the bracket board should not be painted; instead, it can be coated with red lacquer. The base color of the bracket board is usually red, with a green border and a gold inner frame line. Other frame lines may be black or other colors.

Color paintings on rafters. Color painting on rafters refers to the color painting done on the eaves beams and flying beams in Qing Dynasty architecture. The color paintings are located on the body and head of the beams. The body of the beam is only colored, usually in shades of green and blue, without any additional patterns. The patterns used on the eaves beam heads and flying beam heads are different. The eaves beam head patterns typically include dragon eyes, flowers, longevity signs, etc. The base color is usually green, and the borders are decorated with powdered gold. On the other hand, the flying beam head patterns often include the character "万" (meaning "ten thousand"), gardenia flowers, colored flowers, etc. The base color is green, and the borders are also decorated with powdered gold.

Color paintings on ceilings. Color paintings on ceilings in the Qing Dynasty can be divided into three parts: ceiling coloring, wellhead panel patterns, and support beam patterns. Ceiling coloring includes circular light, square light, wide border, supporting beams, and border lines. Circular light is colored in green or blue, with gold or red border lines. Square light is colored in light green

TOP Color paintings on rafters.

BOTTOM Color paintings on ceilings.

or light blue, with gold or red border lines. The wide outline is dark green. Wellhead lines and support beam border lines are colored in gold. The patterns on the wellhead panel usually consist of a central drum with patterns such as two dragons playing with a pearl, two phoenixes facing the sun, two cranes dancing, five bats holding longevity peaches, and golden lotus flowers. The border lines of support beams, also known as wellhead lines, are usually decorated with gold or colored lines. The patterns used on the support beams include clouds, *kui* dragons, flowers, and grass. The tail end of these patterns resembles a double-forked swallowtail, so they are also called cloud swallowtail, *kui* dragon swallowtail, flower swallowtail, and grass swallowtail. The specific type of support beam swallowtail used should be coordinated with the pattern on the wellhead board.

NOTES

1. This beautiful consciousness, which is the embryonic form of aesthetic consciousness, includes the early adornment of the body and objects, and of course, these decorations are inevitably based on functionality. It is these embryonic forms of aesthetics that have led our ancestors to develop a sense of decoration in the process of architectural construction.
2. Yang Hongxun, *Chinese Cultural Studies Journal*, vol. 1 (Shanghai: Fudan University Press, 1984), 66.
3. Chen Yanqing, *The Color of Ideas: Research on Traditional Chinese Colors* (Beijing: Peking University Press, 2015), 7.
4. In the traditional five-element colors, the names do not correspond to the color names in the modern scientific sense. For example, in modern color concepts, "qing" includes the colors green, blue, and cyan; "feihong" and "zhudan" both refer to red; "tuhuang," "cihuang," and "ganse" all represent yellow.
5. Pan Guxi, *The History of Chinese Architecture* (Beijing: China Architecture & Building Press, 2015), 293.
6. Li Yunhe, *Principles of Classical Chinese Architectural Design* (Tianjin: Tianjin University Press, 2005), 279.
7. Wikipedia: A continuous pattern formed by arranging one or more unit patterns in a regular manner on a strip plane between two parallel lines, and continuously repeating in an infinite loop in the upward, downward, or horizontal direction is called a two-way continuous pattern.
8. Tian Yongfu, *Handbook of Knowledge on Ancient Chinese Architecture* (Beijing: China Architecture & Building Press, 2013), 408.

AFTERWORD

The vast and intricate realm of ancient Chinese architecture encompasses a myriad of categories, including residential dwellings, imperial palaces, ceremonial structures, religious edifices, mausoleums, gardens, and utilitarian facilities. Each category forms a distinct system worthy of extensive literary exploration.

It is worth noting that, apart from certain functional structures and facilities, ancient Chinese architecture did not originally classify buildings based on their functionality. Houses were distinguished solely by their hierarchical status, with no differentiation based on their intended use.

The architecture of Buddhist temples is essentially residential architecture. In terms of individual building structures, there is almost no difference from ordinary residential buildings, and even palace architecture is only at a higher level, with no significant changes in its structure. The design principle of traditional Chinese architecture is that a house is a house, regardless of its purpose, and it is hoped to be adaptable. In contrast, Western architecture is driven by its functionality, and buildings such as churches are difficult to be used as residences.

In addition, as stated in the preface of this book, my aspiration is to open the doors for the general public to appreciate the beauty of ancient Chinese architecture, thereby fostering a genuine understanding and recognition of its beauty. Attempting to provide comprehensive discourse on each category would inevitably result in a superficial exploration.

Based on the aforementioned reasons, this book focuses on timber architecture as its point of entry, allowing for a more in-depth exploration. With careful attention, particularly to the section on timber frameworks, readers will gain insight into a better understanding and appreciation of the art of ancient Chinese timber architecture.

The majority of the photographs selected for this book were taken during my years of exploration of ancient buildings. Despite the exhaustion and hardship I experienced when climbing mountains and crossing ridges, facing wind and rain, frost, and snow, I am filled with contentment. Whenever I recall walking alone in remote and rarely visiting deep mountains, the joys and hardships experienced there are particularly profound. This manuscript is the result of compiling and reevaluating the field research conducted in recent years. The manuscript was written during a period of being confined at home due to the global pandemic, where an atmosphere of unease prevailed and the anxiety of writing often intertwined with it. The cyclical year of Gengzi, marking sixty years, is destined to be extraordinary. The global pandemic continues to spread, accompanied by over a month of devastating floods in the southern regions, and the path to economic recovery is fraught with challenges. Nevertheless, I remain optimistic that the world will eventually progress and everything will become better.

Liu Haibo
July 2020

ILLUSTRATION CREDITS

All illustrations in this book, except those credited in the Illustration Credits, are the author's own work.

CHAPTER 1

P. 7: Li Yunhe, *Principles of Classical Chinese Architectural Design* (Tianjin: Tianjin University Press, 2005), 12.

P. 10 (left): Ito Chuta, *A Journey through China: Ito Chuta's Architectural Survey Notes* (Beijing: China Pictorial Publishing House, 2017), 7.

P. 10 (right): Zhu Tao, *Liang Sicheng and His Era* (Guilin: Guangxi Normal University Press, 2014), 32.

P. 17: Photo by Rou He.

P. 20, P. 27: Wang Guixiang, *Ancient Chinese Dwelling Concepts and Architectural Principles* (Beijing: China Architecture & Building Press, 2015), 32, 120, 178.

CHAPTER 2

P. 40, P. 43: Yang Hongxun, *Collection of Architectural Archaeology Papers by Yang Hongxun (Revised Edition)* (Beijing: Tsinghua University Press, 2008), 22, 37.

P. 42: *Archaeology Journal*, no. 1 (1978).

P. 46 (bottom), P. 49, P. 53, P. 59 (bottom): Liu Xujie, *The History of Ancient Chinese Architecture*, vol. 1 (Beijing: China Architecture & Building Press, 2009), 123, 233, 430, 636.

P. 48 (top left): Li Yongdi and Feng Zhongmei, *Selected Excavation Photos of the Yin Ruins 1928–1937* (Taipei: Institute of History and Philology, Academia Sinica, 2012), 22.

P. 48 (top right): *Archaeology*, no. 10 (1989).

P. 50, P. 52 (top), P. 54, P. 56, P. 58, P. 65 (top): Liu Dunzhen, *The Complete Works of Liu Dunzhen*, vol. 9 (Beijing: China Architecture & Building Press, 2007), 21, 20, 68, 69, 70, 71, 44, 90.

P. 65 (bottom): Charles Lang Freer, *Endless Light of Buddha: Freer's 1910 Longmen Journey* (Shanghai: Shanghai Calligraphy and Painting Publishing House, 2014), 203.

P. 66: Hou Youbin and Li Wanzhen, *Illustrated History of Ancient Chinese Architecture* (Beijing: China Architecture & Building Press, 2002), 41, 43.

P. 70 (bottom right): Liang Sicheng, *Liang* (Beijing: China Youth Publishing Group, 2013).

CHAPTER 3

P. 117: Drawing by Liang Sicheng.

P. 121 (top), P. 132: Pan Guxi, *The History of Chinese Architecture*, 7th ed. (Beijing: China Architecture & Building Press, 2015), 264, 57.

P. 121 (bottom), P. 121 (middle), P. 138, P. 142: Hou Youbin, *Foundation* (Beijing: China Architecture & Building Press, 2016), 22, 55, 73, 70.

P. 125 (top), P. 127, P. 135, P. 139, P. 141, P. 143, P. 148, P. 152: Liu Dake, *Construction Methods of Ancient Chinese Buildings with Tiles and Stones*, 2nd ed. (Beijing: China Architecture & Building Press, 2015), 20, 18, 413, 414, 319, 417, 406, 407, 201.

P. 145, P. 147: Liu Dunzhen, *The Complete Works of Liu Dunzhen*, vol. 6 (Beijing: China Architecture & Building Press, 2007), 178, 179.

CHAPTER 4

P. 160, P. 163 (top), P. 191, P. 205, P. 219, P. 229, P. 250, P. 263: Liu Dunzhen, *The Complete Works of Liu Dunzhen*, vol. 6 (Beijing: China Architecture & Building Press, 2007), 166, 167, 187, 188, 190, 191, 192, 197, 204.

P. 161, P. 162: Architecture History Group of Nanjing Institute of Technology, *Illustrated History of Chinese Architecture* (1978), 52, 53.

P. 163 (bottom), P. 180, P. 222: Liang Sicheng, *A Pictorial History of Chinese Architecture* (Beijing: China Architecture & Building Press, 1991), 167, 187, 188.

P. 165 (top), P. 173 (top), P. 174 (top left), P. 178 (top): Liu Dunzhen, *The History of Ancient Chinese Architecture* (Beijing: China Architecture & Building Press, 1980), 252, 255.

P. 166 (top), P. 168 (left), P. 169 (right), P. 185 (left), P. 186 (left), P. 241 (top left): Guo Daiheng, *The History of Ancient Chinese Architecture*, vol. 3 (Beijing: China Architecture & Building Press, 2009), 400, 659, 157, 267, 635.

P. 166: Liu Xujie, *The History of Ancient Chinese Architecture*, vol. 1 (Beijing: China Architecture & Building Press, 2009), 160.

P. 168 (right), P. 169 (left), P. 233: Chen Mingda, *Interpretation of "Yingzao Fashi"* (Tianjin: Tianjin University Press, 2010), 52, 468, 32.

P. 170, P. 171 (top), P. 185 (right): Institute for the History of Natural Sciences, Chinese Academy of Sciences, *History of Ancient Chinese Architectural Technology* (Beijing: Science Press, 1985), 196, 208, 158, 180.

P. 172: Hou Youbin, *Foundation* (Beijing: China Architecture & Building Press, 2016), 21.

P. 174 (bottom left): Lei Dongxia, *Illustrated Guide to Classical Chinese Architecture* (Shanghai: Tongji University Press, 2015), 16.

P. 175 (bottom): Liu Dake, *Construction Methods of Ancient Chinese Buildings with Tiles and Stones*, 2nd ed. (Beijing: China Architecture & Building Press, 2015), 286.

P. 182 (bottom): Ma Bingjian, *Construction Techniques of Traditional Chinese Wooden Architecture*, 2nd ed. (Beijing: Science Press, 2003), 6.

P. 186 (right): Fu Xinian, *The History of Ancient Chinese Architecture*, vol. 2 (Beijing: China Architecture & Building Press, 2009), 497.

P. 207: Li Yunhe, *Principles of Classical Chinese Architectural Design* (Tianjin: Tianjin University Press, 2005), 246.

P. 223: Xie Yuming, *Detailed Design of Traditional Chinese Architecture* (Beijing: China Architecture & Building Press, 2001), 118.

P. 225, P. 239 (top), P. 241 (top right), P. 244 (top): Zhang Shiqing, *Zen Buddhist Temple Architecture in Jiangnan, China* (Wuhan: Hubei Education Press, 2002), 177, 180, 192, 181.

P. 237 (top), P. 239 (bottom): Pan Guxi and He Jianzhong, *"Yingzao Fashi" Interpretation* (Nanjing: Southeast University Press, 2005), 66, 89.

P. 240, P. 243 (top left), P. 256 (top): Liang Sicheng, *Annotations on "Yingzao Fashi"* (Beijing: China Architecture & Building Press, 1983), 243, 105, 257, 108, 148.

P. 245, P. 253, P. 258: Zhang Jiaji, *Discussions on Chinese Architecture* (Taiyuan: Shanxi People's Publishing House, 2003), 66, 403, 105.

P. 254 (middle and bottom), P. 255 (top): Tian Yongfu, *Handbook of Knowledge on Ancient Chinese Architecture* (Beijing: China Architecture & Building Press, 2013), 138, 131, 136.

P. 255 (bottom): Zhang Zhongyi, Cao Jianbin, Fu Gaojie, and Du Xiujun, *Ming Dynasty Residences in Huizhou* (Beijing: Building Engineering Press, 1957), 24.

CHAPTER 5

P. 276, P. 277 (bottom), P. 278, P. 279, P. 287 (bottom), P. 288, P. 295 (top), P. 296: Tian Yongfu, *Handbook of Knowledge on Ancient Chinese Architecture* (Beijing: China Architecture & Building Press, 2013), 227, 228, 238, 151.

P. 280 (bottom), P. 282, P. 283, P. 299: Liu Dake, *Construction Methods of Ancient Chinese Buildings with Tiles and Stones*, 2nd ed. (Beijing: China Architecture & Building Press, 2015), 136, 137, 73, 79, 80.

P. 286, P. 287 (top), P. 289, P. 290: Lei Dongxia, *Illustrated Guide to Classical Chinese Architecture* (Shanghai: Tongji University Press, 2015), 124, 122.

CHAPTER 6

P. 305: Liu Dunzhen, *The Complete Works of Liu Dunzhen*, vol. 6 (Beijing: China Architecture & Building Press, 2007), 165.

P. 306, P. 309 (top), P. 310 (top), P. 311 (top), P. 313, P. 314, P. 317, P. 320 (top), P. 323, P. 324, P. 341 (top): Tian Yongfu, *Handbook of Knowledge on Ancient Chinese Architecture* (Beijing: China Architecture & Building Press, 2013), 167, 168, 169, 200, 202, 179, 187, 37, 198.

P. 308, P. 326, P. 330, P. 331, P. 339: Zhang Jiaji, *Discussions on Chinese Architecture* (Taiyuan: Shanxi People's Publishing House, 2003), 325, 332, 330, 336, 324, 337.

P. 334: Institute for the History of Natural Sciences, Chinese Academy of Sciences, *History of Ancient Chinese Architectural Technology* (Beijing: Science Press, 1985), 153, 173.

P. 340: Liu Dake, *Construction Methods of Ancient Chinese Buildings with Tiles and Stones*, 2nd ed. (Beijing: China Architecture & Building Press, 2015), 243.

P. 341 (bottom): Architecture History Group of Nanjing Institute of Technology, *Illustrated History of Chinese Architecture* (1978), 153.

CHAPTER 7

P. 353, P. 378, P. 380, P. 383: Liu Dunzhen, *The Complete Works of Liu Dunzhen*, vol. 6 (Beijing: China Architecture & Building Press, 2007), 212, 215, 217, 216.

P. 359: Chen Mingda, *Interpretation of "Yingzao Fashi"* (Tianjin: Tianjin University Press, 2010), 217.

P. 360, P. 362, P. 363 (bottom), P. 369 (top), P. 375, P. 376 (right): Tian Yongfu, *Handbook of Knowledge on Ancient Chinese Architecture* (Beijing: China Architecture & Building Press, 2013), 325, 330, 331, 386, 385, 384.

P. 364 (left), P. 368 (right): Institute for the History of Natural Sciences, Chinese Academy of Sciences, *History of Ancient Chinese Architectural Technology*, vol. 1 (Beijing: Science Press, 1985), 249, 255.

P. 364 (right), P. 384: Lei Dongxia, *Illustrated Guide to Classical Chinese Architecture* (Shanghai: Tongji University Press, 2015), 136, 159.

P. 367: Guo Daiheng, *The History of Ancient Chinese Architecture*, vol. 3 (Beijing: China Architecture & Building Press, 2009), 688.

P. 370: Pan Guxi, *The History of Chinese Architecture*, 4th ed. (Beijing: China Architecture & Building Press, 2001), 295.

P. 374 (left), P. 382: Ma Bingjian, *Construction Techniques of Traditional Chinese Wooden Architecture* (Beijing: Science Press, 2003), 307, 308.

P. 374 (right): Sun Dazhang, *History of Ancient Chinese Architecture* (Volume Five). Beijing: China Architecture & Building Press, 2009: 482.

P. 376 (left): Zhang Jiaji, *Discussions on Chinese Architecture* (Taiyuan: Shanxi People's Publishing House, 2003), 478.

CHAPTER 8

P. 395: Li Yunhe, *Principles of Classical Chinese Architectural Design* (Tianjin: Tianjin University Press, 2005), 279.

P. 398, P. 400: Pan Guxi and He Jianzhong, *"Ying Zao Fa Shi" Interpretation* (Nanjing: Southeast University Press, 2005), 171, 176, 179, 181, 184, 186.

P. 401, P. 402: Tian Yongfu, *Handbook of Knowledge on Ancient Chinese Architecture* (Beijing: China Architecture & Building Press, 2013), 402.

INDEX